Graphics Tools—

The jgt Editors' Choice

journal of graphics tools

Editor-in-Chief
Ronen Barzel

Consulting Editor
David Salesin

Founding Editor
Andrew Glassner

Editorial Board
Tomas Akenine-Moller
Richard Chuang
Eric Haines
Chris Hecker
John Hughes
Darwyn Peachey
Paul Strauss
Wolfgang Sturzlinger

Graphics Tools—
The jgt Editors' Choice

Ronen Barzel, Editor

A K Peters
Wellesley, Massachusetts

Editorial, Sales, and Customer Service Office

A K Peters, Ltd.
888 Worcester Street, Suite 230
Wellesley, MA 02482
www.akpeters.com

Library of Congress Cataloging-in-Publication Data

Graphics Tools—the jgt editors' choice / Ronen Barzel, editor.-- 1st ed.
   p. cm.
  Includes bibliographical references.
  ISBN 1-56881-246-9
  1. Computer graphics. I. Title: Graphics tools, the Journal of graphics
tools editors' choice. II. Barzel, Ronen. III. Journal of graphics tools.
T385.G734 2005
006.6--dc22

2005048887

Cover image: Figure 3, page 197 by Charles M. Schmidt and Brian Budge.

Printed in the United States of America
09 08 07 06 05   10 9 8 7 6 5 4 3 2 1

Contents

Contents

Contents

Foreword

Happy Anniversary! The *journal of graphics tools* (affectionately known as *jgt*) is celebrating its tenth birthday. We're marking the event with this collection of many of our most useful and interesting articles.

We started *jgt* because as members of the computer graphics community we needed a place to easily share our tools and insights on a regular basis. Existing journals and conferences were great for big research results, and after five volumes, the *Graphics Gems* series of books had run its course as an annual (or biennial) publication. We needed a new home for high-quality, useful, trustworthy, and timely ideas that were too small for a conference but too big to be limited to word of mouth. And thus, *jgt* was born.

Why graphics *tools*? Tools are essential to creative work. We all use tools, every day, even if we never come near a computer. Our tools shape the way we see the world and influence how we act.

When I was in graduate school I owned a car with two gearshifts. One controlled the standard 5-speed manual transmission, while the other was basically a two-position switch that shifted the car from front-wheel drive to 4-wheel drive. At that time very few passenger cars had 4-wheel drive, and friends often wanted to know if I ever really needed to use it. I enjoyed telling them about some sticky situations where it was a lifesaver: stuck rim-deep in soft sand at the beach, climbing through a snowdrift in the mountains, or pushing through deep mud coming home from a hiking trip. Then one day someone observed that I wouldn't have gotten into most of those situations in the first place if I didn't know that I had the 4-wheel drive available to get me out. A light bulb went off over my head. I realized that with the purchase of this car I had unconsciously changed my very idea of where it was safe to drive. Having this tool at my disposal changed my behavior.

Whether we're building a bookshelf, driving in a blizzard, or writing a new animation program, our tools shape the ideas we have, the techniques we employ to realize them, and the artifacts we ultimately create. In computer graphics we have every bit as great a need for the best possible tools as in any other creative field.

Sometimes a tool doesn't look revolutionary on the face of it. I know I've seen tools, both hardware and software, that have struck me as merely an easier way to do what I could already do. But that attitude can be deceptive. Consider the difference between a hand saw and a power saw. If you're willing to invest enough time and energy, you can walk into a tree-filled glade with nothing but an axe and a hand saw and build a house. But with a power saw you can cut more wood than with the hand tool and do it more quickly and more accurately. Naturally, this will make it easier to tackle more ambitious and complex projects. Theoretically, a hand saw and a power saw do the same thing, but in practice, the more powerful tool lets you take on projects that you wouldn't otherwise even consider.

Our tools influence our work. As the tools improve, some ideas move from simply possible to practical, or even convenient. The best tools are simple and robust. We can use them the way they were intended, or adapt them for purposes their designers never imagined. We can do this because great tools have clear limits: we know how far we can push them before they break.

A new tool may be simple and easy to learn, but its best use may not be obvious. On a recent trip to a hardware store I saw at least a half dozen different types of screw heads on one rack alone (slotted, Phillips, Robertson or square, hex, one-way, and TORX). They're all screw heads and all do approximately the same thing, but there's a reason for these different shapes. A skilled carpenter understands the differences and chooses the right screw for each job.

It's something of a folk theorem among tool developers that users are not the best source of ideas for new developments. A tool user focuses on results and only notices a tool when it fails or surprises in some way (as I write these words, I'm oblivious to my text editor, operating system, and keyboard). A tool designer focuses on process and looks for ways to simplify or enhance the user's experience or capabilities.

As our tools grow in sophistication, they free us to dream and imagine at ever higher conceptual and abstract levels, resulting in more ideas and faster experimentation.

These pages present some of the best of the many reliable, useful tools we've published. The papers in this book demonstrate the wide variety of creative discovery that continues to bring people to computer graphics.

May you find that these tools inspire you to take on new projects, explore new territory, and in the course build new tools that you'll share with everyone else through the pages of *jgt*!

Andrew Glassner
Founding Editor *jgt*

Preface

The *journal of graphics tools* is now in its tenth year. Since the start, our mission has been to provide useful tools for the computer graphics practitioner—and at this point we have published quite a few. But in a field as quickly changing as ours, the utility of a tool has a somewhat ephemeral quality. It's easy to be suspicious of techniques even just a few years old. Are they still relevant? Has something better come along? This book is a response to those concerns: an attempt to gather together some of the most useful and still-relevant techniques that have appeared in *jgt* over the years.

So here we have, in time to celebrate our tenth anniversary, the "editors' choice" collection of papers, drawn from the first nine volumes of *jgt*. But how did the editors choose? We have no single well-defined dimension of quality; everyone has their own personal opinions and metrics. So we asked each member of the editorial board to compile a list of personal favorites using the criteria they personally thought appropriate: tools that they use often, ideas that are particularly clever, information that's still not easily available elsewhere, and so forth. When we put the resulting lists together, we found ourselves with uncomfortably more papers than would fit in a single volume. Of the many papers that we wished we could include, we limited ourselves to those that appeared on several editors' personal lists, and we tended towards shorter papers in order to include as many as possible.

Reprinting these papers gave us the opportunity to invite authors to add follow-up comments discussing new developments since their original work. These notes appear at the end of each paper in a section titled "New Since Original Publication." We've also updated the authors' contact information. And, of course, we corrected known typos and errata. Aside from that, the papers are printed here in their original form.

For those of you who are long-time subscribers to *jgt,* I hope that this book will serve as a useful reference. And for those of you who haven't seen the journal—now you know what you've been missing! I invite you to subscribe and keep up with the tools as they emerge.

And for those of you who read one of these papers and think, "They published *that*?! I know a much better way to do it!"—I certainly encourage you to write it up and submit it to the journal! I can think of no better way for this book to provide a useful service to the community than by inspiring the community to render it obsolete.

It's customary in a preface for the author of a book to thank everyone who helped make it possible. Of course I'm not the author of this book, only of this preface. But still, I'd like to acknowledge those whose hard work and dedication made not only this book a reality, but the last ten years of *jgt*: Alice Peters and the staff at A K Peters, without whom none of this would happen; Andrew Glassner, who created *jgt* and has advised us all along; the editorial board: Tomas Akenine-Möller, Richard Chuang, Eric Haines, Chris Hecker, John Hughes, Darwyn Peachey, David Salesin, Peter Shirley, Paul Strauss, and Wolfgang Stürzlinger; and the anonymous reviewers. Of course, *jgt* would be nothing without the authors who have contributed their work over the years—it's the ongoing work of the members of the community and their willingness and eagerness to share their knowledge, that provide the benefit to us all. And finally, thanks to the authors of the papers in this collection for taking the time to revisit their work and prepare it for republication.

Ronen Barzel

Part I

Math and Programming Techniques

Vol. 7, No. 3: 1–11

Simple and Efficient Traversal Methods for Quadtrees and Octrees

Sarah F. Frisken and Ronald N. Perry

Mitsubishi Electric Research Laboratories

Abstract. Quadtrees and octrees are used extensively throughout computer graphics and in many other diverse fields such as computer vision, robotics, and pattern recognition. Managing information stored in quadtrees and octrees requires basic tree traversal operations such as point location, region location, and neighbor searches. This paper presents simple and efficient methods for performing these operations that are inherently nonrecursive and reduce the number of comparisons with poor predictive behavior. The methods are table-free, thereby reducing memory accesses, and generalize easily to higher dimensions. Source code is available online.

1. Introduction

Quadtrees and octrees are spatial data structures that successively partition a region of space into four or eight equally sized quadrants or octants (i.e., cells). Starting from a root cell, cells are successively subdivided into smaller cells under certain conditions, such as when a cell contains an object boundary (e.g., region quadtree) or when a cell contains more than a specified number of objects (e.g., point quadtree). Compared to methods that do not partition space or that partition space uniformly, quadtrees and octrees can reduce the amount of memory required to represent objects (e.g., an image) and improve execution times for querying and processing data (e.g., collision detection).

Quadtrees and octrees can be represented in a hierarchical tree structure composed of a root cell, intermediate cells, and leaf cells, or in a pointerless representation such as a linear quadtree, which stores only leaf cells in a cell list. In tree-based representations, each cell stores pointers to its parent and child cells. Operations such as point location (finding the leaf cell containing a given point) and neighbor searches (finding a cell in a specified direction that touches a given cell) traverse the tree by following these pointers. In linear quadtrees, each cell has an associated *locational code*, which acts as a search key to locate the cell in the cell list. Traditionally, these locational codes interleave bits that comprise values of the cell's minimum $(x, y, z$, etc.) coordinates such that quadtrees use locational codes of base 4 (or 5 if a "don't care" directional code is used) and octrees use locational codes of base 8 (or 9) (see [Samet 90b]). In general, linear quadtrees are more compact than tree-based representations at the expense of more costly or complicated processing methods.

Managing information stored in a quadtree or octree requires basic operations such as point location, region location (finding the smallest cell or set of cells that encloses a specified region), and neighbor searches. Traditional methods for point location in a tree-based representation require a downward branching through the tree from the root node, where each branch is made by comparing a point's position to the midplane positions of the current enclosing cell. In linear quadtrees, point location is performed by encoding a point's position into a locational code and then searching the (ordered) cell list to find the cell whose locational code best matches that of the point position. Traditional neighbor searching in a tree-based representation (e.g., [Samet 90b]) uses a recursive upward branching from the given cell to the smallest common ancestor of the cell and its neighbor, and then a recursive downward branching to locate the neighbor. Each branch in the recursion relies on comparing values (derived from tables) that depend on the current cell and its parent. Neighbor searches in linear quadtrees can be performed in a similar manner [Samet 90b].

As discussed in [Knuth 98] and [Knuth 99], the comparisons required for point location, region location, and neighbor searching can be costly because the predictive branching strategies used for modern CPUs will stall the instruction pipeline whenever branch instructions are incorrectly predicted. Such mispredictions occur frequently during tree traversal because, unlike most loops, the branch chosen in a previous iteration has no relevance to the likelihood of it being taken again [Pritchard 01]. In addition, traditional neighbor searching methods are recursive (thus introducing overhead from maintaining stack frames and making function calls), table-based (thus requiring costly memory accesses in typical applications), and difficult to extend to higher dimensions. Here, we present simple and efficient tree traversal methods for tree-based representations of quadtrees, octrees, and their higher-dimensional counterparts that use locational codes and which are inherently nonrecursive, reduce or eliminate the number of comparisons with poor predictive behavior, and are table-free. We provide details of the representation, algorithms for point location, region location, and neighbor searching, and C code (at the web site listed at the end of this paper) that performs these operations. While we focus on quadtrees, extension to octrees and higher-dimensional trees is trivial because these methods treat each dimension independently.

2. Representation

We assume a standard tree-based quadtree representation of a cell hierarchy branching from a root cell, through intermediate cells, to leaf cells. Each cell contains a pointer to its parent cell (null for the root cell), a pointer to the first of four consecutively stored child cells (null for leaf cells), and (typically) an application-specific field for cell data (e.g., a cell type for region quadtrees or object indices for point quadtrees). The depth of the quadtree is limited to N_LEVELS for practical purposes. Similar to [Samet 90a], we define the level of the root cell to be ROOT_LEVEL $\equiv$ (N_LEVELS - 1), and the level of the smallest possible cell in the tree to be 0. The quadtree is defined over $[0, 1] \times [0, 1]$, which can be achieved for a general rectangular region by applying an affine transformation of the coordinate space.

A locational code is associated with each coordinate of a cell's minimum vertex and is represented in binary form in a data field with bit size greater than or equal to N_LEVELS. Unlike traditional locational codes, we do not interleave bits for x- and y-coordinates into a single value and instead use independent locational codes for each coordinate. The bits in a locational code are numbered from right (LSB) to left starting from 0. Each bit in a locational code indicates the branching pattern at the corresponding level of the tree, i.e., bit k represents the branching pattern at level k in the tree.

Figures 1 and 2 illustrate how locational codes are determined for a "bi-tree," the one-dimensional equivalent of a quadtree. Figure 1 is a spatial representation of the bi-tree which is defined over [0,1]. The bi-tree has six levels (i.e., N_LEVELS = 6) and the level of the root cell is 5 (i.e., ROOT_LEVEL = 5). Figure 2 illustrates the same spatial partitioning with a tree representation.

Position

| 0.00 | 0.125 | 0.25 | 0.50 | 0.625 | | 0.75 | 1.00 |

| 000000 | 000100 | 001000 | 010000 | 010100 | | 011000 |
| | | | | | 010110 | 010111 |

Locational code

Figure 1. A one-dimensional spatial partitioning over $[0, 1]$ showing the relationship between the position of the cell's left corner and the cell's locational code.

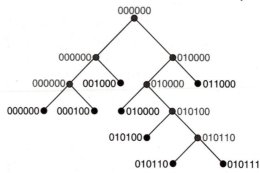

Figure 2. The same spatial partitioning of Figure 1 represented as a binary tree showing the relationship between a cell's locational code and the branching pattern from the root cell to the cell.

The locational code for each cell can be determined in two ways. The first method multiplies the position of the cell's left corner by $2^{\text{ROOT\_LEVEL}}$ (i.e., $2^5 = 32$) and then represents the product in binary form. For example, the cell $[0.25, 0.5)$ has locational code $binary(0.25 * 32) = binary(8) = 001000$. The second method follows the branching pattern from the root cell to the cell in question, setting each bit according to the branching pattern of the corresponding level. Starting by setting bit ROOT_LEVEL to 0, the method then sets each subsequent bit, k, to 0 if the branching pattern from level $k + 1$ to k branches to the left and to 1 if it branches to the right. All lower order bits that are not set because a leaf cell is reached are set to 0. In quadtrees (octrees, etc.), locational codes for each dimension are computed independently from the x, y, (z, etc.) positions of the left, bottom, (back, etc.) cell corner (first method) or from the left-right, bottom-top, (back-front, etc.) branching patterns used to reach the cell in question from the root cell (second method).

Several properties of these locational codes can be used to devise simple and efficient tree traversal methods. First, just as locational codes can be determined from branching patterns, branching patterns can be determined from locational codes. That is, a cell's locational code can be used to traverse the tree (without costly comparisons) from the root cell to the cell by using the appropriate bit in the x, y, (z, etc.) locational codes to index the corresponding child of each intermediate cell. Second, the position of any point in $[0, 1) \times [0, 1)$ ($[0, 1)^3$, etc.) can be converted into 2 (3, etc.) locational codes by applying the first method for determining cell locational codes to each coordinate of the point. These first two properties enable efficient point and region location as described in Sections 3 and 4. Finally, the locational codes of any neighbor of a cell can be determined by adding and subtracting bit patterns to the cell's locational codes. This property is used in Section 5 to provide simple, nonrecursive, table-free neighbor searches that generalize easily to higher dimensions. This property is also used in Section 6 to provide an efficient, bottom-up method for determining leaf cells along a ray intersecting an octree (an operation required for ray tracing objects stored in an octree).

3. Point Location

For point location (i.e., determining a leaf cell that contains a given point), we assume that the quadtree represents the region $[0, 1] \times [0, 1]$ and the point lies in $[0, 1) \times [0, 1)$. The first step converts the point's x- and y-coordinate values to x and y locational codes by multiplying each coordinate value by $2^{\text{ROOT\_LEVEL}}$, truncating the resultant products to integer types, and representing the integers in binary form. Assuming that floating point numbers are represented using the IEEE standard, this conversion to integers can be performed very efficiently [King 01]. The second step starts at the root cell, and at each level k in the tree, uses the $(k-1)^{\text{st}}$ bit from each of the x, y, (z, etc.) locational codes to determine an index to the appropriate child cell. Note that either all of the cell's four (eight, etc.) children or all of the pointers to these children are stored consecutively and are consistently ordered to enable this indexing. When the indexed child cell has no children, the desired leaf cell has been reached and the operation is complete.

As an example, in the bi-tree of Figures 1 and 2, the point at $x = 0.55$ has locational code $binary(trunc(0.55 * 32)) = binary(17) = 010001$. Following the branching pattern of the point's locational code from the root cell in Figure 2 (i.e., branching first right, then left, then left where a leaf cell is reached), the point is located in the leaf cell $[0.5, 0.625)$ which has locational code 010000.

Like traditional methods for point location, this method requires a comparison at each level of the tree to test if the current cell is a leaf cell. For example, in order to locate a point in a level 0 cell of an 8-level octree, both methods require 8 of these simple leaf-cell tests. However, unlike traditional methods, comparisons between the point position and midplane positions of each cell are *not* needed at each branching point. This saves d comparisons at each level in the tree traversal, where d is the dimension of the tree (e.g., 2 for quadtrees, 3 for octrees, etc.). Therefore, in order to locate a point in a level 0 cell of an 8-level octree, traditional methods require an additional 24 ($= 3 * 8$) comparisons to branch to the appropriate children of intermediate cells. These additional comparisons in traditional methods exhibit the poor predictive behavior described by [Pritchard 01].

4. Region Location

Region location involves finding the smallest cell or set of cells that encloses a given region. Here we assume that the region is a rectangular, axis-aligned bounding box and present a method for finding the single smallest cell entirely enclosing the region.

[Pritchard 01] recently presented a method for region location in quadtrees that uses binary representations (i.e., locational codes) of the x and y boundaries of the region bounding box. However, his method assumes a quadtree representation that is not truly tree-based. Instead, his quadtrees are represented by a hierarchy of regular arrays of cells, where each level is fully subdivided and contains four times as many cells as the previous level. This representation alleviates the need to store pointers in each cell and allows simple indexing for cell access at each level. However, it requires significantly more memory than tree-based quadtrees (and even more for octrees) and hence, is impractical for many applications. Pritchard's method has two steps: It first uses the locational codes of the left and right x boundaries and the top and bottom y boundaries of the region bounding box to determine the level of the enclosing cell and then uses a scaled version of the position of the bottom-left vertex of the region bounding box to index into the regular array at this level. We provide a method for region location in tree-based quadtrees, octrees, and their higher-dimensional counterparts which first determines the size of the smallest possible enclosing cell using a method similar to Pritchard's first step, and then uses a variation of the point location method presented above to traverse the tree from the root cell to the smallest enclosing cell.

Similar to Pritchard, we determine the size (i.e., level) of the smallest possible enclosing cell by XORing the left and right x locational codes and the top and bottom y locational codes of the left-right and top-bottom vertices of the region. These two XORed codes are then searched from the left (MSB) to find the first 1 bit of the two codes, indicating the first level below the root level where at least one of the pairs of locational codes differ. The level of the smallest possible enclosing cell is then equal to the bit number of the 0 bit immediately preceding this 1 bit. Given this level, our method then traverses the tree downward from the root cell following the bit pattern of the locational codes of any point in the region (e.g., the bottom-left corner) until a leaf cell is encountered or a cell of the determined size is reached. This yields the desired enclosing cell. Note that there are several methods for identifying the highest order 1 bit in the XORed values ranging from a simple shift loop to platform-specific single instructions which bit-scan a value, thereby eliminating the loop and subsequent comparisons.

As a first example, using the bi-tree of Figures 1 and 2, the region [0.31, 0.65) has left and right locational codes 001001 and 010101, respectively. XORing these locational codes yields 011100, with the first 1 bit from the left (MSB) encountered at bit 4, so that the level of the smallest possible enclosing cell is 5, i.e., the enclosing cell of the region [0.31, 0.65) is the root cell. As a second example, the region [0.31, 0.36) has locational codes 001001 and 001010. The XOR step yields 000011, with the first 1 bit from the left encountered at bit 1, so that the level of the smallest possible enclosing cell is 2. The smallest enclosing cell can then be found by traversing the bi-tree downward from the root cell following the left locational code, 001001, until a leaf cell is encountered or a level 2 cell is reached. For the spatial partitioning of Figure 1, the enclosing cell is the level 3 leaf cell with locational code 001000.

5. Neighbor Searches

Neighbor searching finds a cell in a specified direction (e.g., left, top, top-left, etc.) that touches a given cell. Several variations exist, including finding vertex, or edge, (or face, etc.) neighbors, finding neighbors of the same size or larger than the given cell, or finding all of the leaf cell neighbors of the given cell.

In order to determine neighbors of a given cell, we first note that the bit patterns of the locational codes of two neighboring cells differ by the binary distance between the two cells. For example, we know that the left boundary of every right neighbor of a cell (including intermediate and leaf cells) is offset from the cell's left boundary by the size of the cell. Hence, the x locational code of every right neighbor of a cell can be determined by adding the binary form of the cell's size to the cell's x locational code. Fortunately, the binary form of a cell's size is easily determined from its level, i.e., `cellSize` $\equiv$ `binary(`$2^{\tt cellLevel}$`)`. Hence, the x locational code for a cell's right neighbor is the sum of the cell's x locational code and `cellSize`.

As an example, the cell [0.25, 0.5) of the bi-tree in Figures 1 and 2 has locational code 001000 and level 3. Hence, the x locational code of a neighbor touching its right boundary is $001000 + binary(2^3) = 001000 + 001000 = 010000$ which can be verified by inspection.

Determining the x locational codes of a cell's left neighbors is more complicated. Because we don't know the sizes of the cell's left neighbors, we don't know the correct binary offset between the cell's x locational code and the x locational codes of its left neighbors. However, we first observe that the smallest possible left neighbor has level 0 and hence the difference between the x locational code of a cell and the x locational code of the cell's smallest possible left neighbor is $binary(2^0)$ (i.e., smallest possible left neighbor's x locational code = cell's x locational code $-$ $binary(1)$). Second, we observe that the left boundary of this smallest possible left neighbor is located between the left and right boundaries of every left neighbor of the cell (including intermediate cells). Hence, a cell's left neighbors can be located by traversing the tree downward from the root cell using the x locational code of this smallest possible left neighbor and stopping when a neighbor cell of a specified level is reached or a leaf cell is encountered.

As an example, referring to Figure 1, the smallest possible left neighbor of the cell [0.25, 0.5) has locational code $001000 - 000001 = 000111$. Traversing the tree downwards from the root cell using this locational code and stopping when a leaf cell is reached yields the cell [0.125, 0.25) with locational code 000100 as the cell's left neighbor.

For quadtrees (octrees, etc.), a neighbor is located by following the branching patterns of the pair (triplet, etc.) of x, y, (z, etc.) locational codes to the neighbor until a leaf cell is encountered or a specified maximum tree traversal level is reached. Each of these locational codes is determined from the appropriate cell boundary (which is determined from the specified direction to the neighbor). In a quadtree, the x locational code of a right edge neighbor is determined from the cell's right boundary and the x and y locational codes of a top-right vertex neighbor are determined from the cell's top and right boundaries.

For example, the right edge neighbor of size greater than or equal to a cell is located by traversing downward from the root cell using the x locational code of the cell's right boundary and the y locational code of the cell until either a leaf cell or a cell of the same level as the given cell is reached. As a second example, a cell's bottom-left leaf cell vertex neighbor is located by traversing the tree using the x locational code of the cell's smallest possible left neighbor and the y locational code of the cell's smallest possible bottom neighbor until a leaf cell is encountered.

Once the locational codes of the desired neighbor have been determined, the desired neighbor can be found by traversing the tree downward from the root cell. However, as described in [Samet 90b], it can be more efficient to first traverse the tree upward from the given cell to the smallest common ancestor of the cell and its neighbor and then traverse the tree downward to the neighbor. Fortunately, locational codes also provide an efficient means for determining this smallest common ancestor.

Assuming a bi-tree, the neighbor's locational code is determined, as described above, from the given cell's locational code and the specified direction to the neighbor. The given cell's locational code is then XORed with the neighbor's locational code to generate a difference code. The bi-tree is then traversed upward from the given cell to the first level where the corresponding bit in the difference code is a 0 bit. This 0 bit indicates the smallest level in the bi-tree where the two locational codes are the same. The cell reached by traversing upward to this level is the smallest common ancestor.

In quadtrees (octrees, etc.), the 2 (3, etc.) locational codes of the given cell are XORed with the corresponding locational codes of the neighbor to produce 2 (3, etc.) difference codes. The cell with the highest level reached by the upward traversal for each of these difference codes is the smallest common ancestor.

As an example, referring to Figure 1, the difference code for the level 3 cell [0.25, 0.5) and its right neighbor is 001000 ∧ 010000 = 011000. Traversing the tree upward from level 3 considers bits in this difference code to the left of bit 3. The first 0 bit is reached at ROOT_LEVEL, so the smallest common ancestor of [0.25, 0.5) and its right neighbor is the root cell. As a second example, the difference code for the level 3 cell [0.75, 1) and its left neighbor is 011000 ∧ 010111 = 001111. Examining bits to the left of bit 3 yields the first 0 at bit 4, corresponding to a level 4 cell. Hence, the smallest common ancestor of the cell [0.75, 1) and its left neighbor is the cell's parent cell [0.5, 1) which has locational code 010000.

Depending on the application, several different variations of neighbor searches might be required. C code is provided at the web site listed at the end of this paper for some typical examples, i.e., finding the smallest left and right neighbors of size at least as large as the given cell and finding the three leaf cell neighbors touching the given cell's bottom-right vertex.

The main advantage of this neighbor finding method over traditional methods is its simplicity: It generalizes easily to higher dimensions and it can be applied to many variations of neighbor searches. In contrast, traditional neighbor searches require different methods for face, edge, and vertex neighbors and "vertex neighbors are considerably more complex" [Samet 90b]. Our method trades off traditional table lookups for simple register-based computations in the form of bit manipulations. This is advantageous in current architectures where CPU speeds far exceed memory speeds. Note that if the working set of an application is fortunate enough to be cache resident, table lookups are virtually free. However, in many practical applications, the application data and the table data compete for the cache, forcing frequent reloading of the table data from memory.

In addition, while [Samet 90b] provides tables for some varieties of neighbor searching, neither he nor the literature provides them for all common variations. Furthermore, tables are specialized for a given cell enumeration and must be redetermined for different cell labeling conventions. Our experience has been that determining these tables requires good spatial intuition and hence they can be error prone, are tedious to verify, and are extremely difficult to generate in 4 dimensions and beyond (where there are numerous applications in fields such as computer vision, scientific visualization, and color science). Finally, our neighbor searching method is inherently nonrecursive and requires fewer comparisons than traditional methods. In contrast, traditional methods for neighbor searching are inherently recursive and unraveling the recursion is nontrivial. [Bhattacharya 01] recently presented a nonrecursive neighbor searching method for quadtrees and octrees limited to finding neighbors of the same size or larger than a cell. However, like Samet's, his method requires table-based tree traversal to determine the appropriate neighbor, and both methods require a similar number of comparisons.

11

6. Ray Traversal

Ray tracing methods often make use of octrees to accelerate tracing rays through large empty regions of space and to minimize the time spent computing ray-surface intersections. These methods determine nonempty leaf cells along a ray passing through the octree and then process ray-surface intersections within these cells. There are two basic approaches for following a ray through an octree: Bottom-up methods which start at the first leaf cell encountered by the ray and then use neighbor-finding techniques to find each subsequent leaf cell along the ray, and top-down methods that start from the root cell and use a recursive procedure to find offspring leaf cells that intersect the ray. An extensive summary of the research in this area is presented in [Havran 99].

Both the top-down and bottom-up approaches can exploit the tree traversal methods described in this paper. In fact, [Stolte, Caubet 95] presents a top-down approach that uses a method similar to the point location method described in Section 3. They first locate a leaf cell containing the point where a ray enters the octree. Then, for each leaf cell without a ray-surface intersection, a three-dimensional DDA is used to incrementally step along the ray (in increments proportional to the size of the smallest possible leaf cell) until a boundary of the leaf cell is crossed, providing a sample point in the next leaf cell along the ray. The next leaf cell is then found by popping cells from a recursion stack to locate a common ancestor of the leaf cell and the next leaf cell and then traversing down the octree using their point location method.

Stolte and Caubet's top-down method can be converted to a nonrecursive bottom-up approach by using the neighbor-searching techniques presented in Section 5. In addition, their method can be improved in two ways. First, the three (dimension-dependent) comparisons in their point location method can be replaced by three multiplies. Second, the use of the three-dimensional DDA can be avoided by using the current leaf cell's size and locational codes to determine one or more of the new locational codes of the next leaf cell directly (from which the entry point to the next leaf cell can be determined).

7.   Conclusions

We have provided methods for point location, region location, and neighbor searching in quadtrees and octrees that are simple, efficient, inherently nonrecursive, and reduce the number of comparisons with poor predictive behavior. The methods are table-free, thereby reducing memory accesses, and generalize easily to higher dimensions. In addition, we have provided C code (at the web site listed at the end of this paper) for point location, region location, and three variations of neighbor searching. In our real-time sculpting application [Perry, Frisken 01], we have found that these methods significantly reduce computation times during triangulation of an octree-based volumetric representation (which requires neighbor searching) and in a time-critical edit-octree/refresh-display loop (which requires point and region location).

References

[Bhattacharya 01] P. Bhattacharya. "Efficient Neighbor Finding Algorithms in Quadtree and Octree." M.T. thesis, Dept. Comp. Science and Eng., India Inst. Technology, 2001.

[Havran 99] V. Havran. "A Summary of Octree Ray Traversal Algorithms." *Ray Tracing News* 12:2 (1999), 11–23.

[King 01] Y. King. "Floating-Point Tricks: Improving Performance with IEEE Floating Point." In *Game Programming Gems 2*, edited by M. DeLoura, pp. 167–181. Hingham, MA: Charles River Media, 2001.

[Knuth 98] D. Knuth. *The Art of Computer Programming*, Vol. 1. Reading, MA: Addison-Wesley, 1998.

[Knuth 99] D. Knuth. *MMIXware: A RISC Computer for the Third Millennium.* New York: Springer-Verlag, 1999.

[Perry, Frisken 01] R. Perry and S. Frisken. "Kizamu: A System for Sculpting Digital Characters." In *Proceedings of SIGGRAPH 2001, Computer Graphics Proceedings, Annual Conference Series*, edited by E. Fiume, pp. 47–56. Reading, MA: Addison-Wesley, 2001.

[Pritchard 01] M. Pritchard. "Direct Access Quadtree Lookup." In *Game Programming Gems 2*, edited M. DeLoura, pp. 394–401. Hingham, MA: Charles River Media, 2001.

[Samet 90a] H. Samet. *The Design and Analysis of Spatial Data Structures.* Reading, MA: Addison-Wesley, 1990.

[Samet 90b] H. Samet. *Applications of Spatial Data Structures: Computer Graphics, Image Processing, GIS.* Reading, MA: Addison-Wesley, 1990.

[Stolte, Caubet 95] N. Stolte and R. Caubet. "Discrete Ray-Tracing of Huge Voxel Spaces." *Computer Graphics Forum* 14:3 (1995), 383–394.

Web Information:

C source code for the point location and region location methods and three example neighbor searches is available at
http://www.acm.org/jgt/papers/FriskenPerry02.

Vol. 7, No. 3: 1–11

Sarah F. Frisken, Mitsubishi Electric Research Laboratories, 201 Broadway, 8th Floor, Cambridge, MA 02139 (frisken@merl.com)

Ronald N. Perry, Mitsubishi Electric Research Laboratories, 201 Broadway, 8th Floor, Cambridge, MA 02139 (perry@merl.com)

Received July 12, 2002; accepted in revised form November 6, 2002.

Current Contact Information:

Sarah F. Frisken, Tufts University, Department of Computer Science, 161 College Ave, Medford, MA 02115 (frisken@cs.tufts.edu)

Ronald N. Perry, Mitsubishi Electric Research Lab, 201 Broadway, Cambridge, MA 02139 (perry@merl.com)

Vol. 4, No. 4: 33–35

Building an Orthonormal Basis from a Unit Vector

John F. Hughes
Brown University

Tomas Möller
Chalmers University of Technology

Abstract. We show how to easily create a right-handed orthonormal basis, given a unit vector, in 2-, 3-, and 4-space.

1. Introduction

Often in graphics, we have a unit vector, $\mathbf{u}$, that we wish to extend to a basis (i.e., we want to enlarge the set $\{\mathbf{u}\}$ by adding new vectors to it until $\{\mathbf{u}, \mathbf{v}, \mathbf{w}, \dots\}$ is a basis, as in, e.g., [Hoffman, Kunze 71], Section 2.5, Theorem 5). For example, when we want to put a coordinate system (e.g., for texture-mapping) on a user-specified plane in 3-space, the natural specification of the plane is to give its normal, but this leaves the choice of plane-basis ambiguous up to a rotation in the plane. We describe the solution to this problem in two, three, and four dimensions.

2. Two Dimensions and Four Dimensions

Two dimensions and four dimensions are the easy cases: To extend $\mathbf{u} = (x, y)$ to an orthonormal basis of $\mathbb{R}^2$, let

$$\mathbf{v} = (-y, x)$$

This corresponds to taking the complex number $x + iy$ and multiplying by i, which rotates 90 degrees clockwise. To extend $\mathbf{u} = (a, b, c, d)$ to an orthonormal basis of $\mathbb{R}^4$, let

$$
\begin{aligned}
\mathbf{v} &= (-b, \quad a, -d, \quad c) \\
\mathbf{w} &= (-c, \quad d, \quad a, -b) \\
\mathbf{x} &= (-d, -c, \quad b, \quad a).
\end{aligned}
$$

This corresponds to multiplying the quaternion $a + bi + cj + dk$ by i, j, and k, respectively.

3. Three Dimensions

Oddly, three dimensions are harder—there is no continuous solution to the problem. If there were, we could take each unit vector $\mathbf{u}$ and extend it to a basis $\mathbf{u}, \mathbf{v}(\mathbf{u}), \mathbf{w}(\mathbf{u})$, where $\mathbf{v}$ is a continuous function. By drawing the vector $\mathbf{v}(\mathbf{u})$ at the tip of the vector $\mathbf{u}$, we would create a continuous nonzero vector field on the sphere, which is impossible [Milnor 65].

Here is a numerically stable and simple way to solve the problem, although it is not continuous in the input: Take the smallest entry (in absolute value) of $\mathbf{u}$ and set it to zero; swap the other two entries and negate the first of them. The resulting vector $\bar{\mathbf{v}}$ is orthogonal to $\mathbf{u}$ and its length is at least $\sqrt{2/3} \approx .82$. Thus, given $\mathbf{u} = (x, y, z)$ let

$$
\begin{aligned}
\bar{\mathbf{v}} &= \begin{cases} (\ \ 0, -z, y), \text{ if } |x| < |y| \text{ and } |x| < |z| \\ (-z, \quad 0, x), \text{ if } |y| < |x| \text{ and } |y| < |z| \\ (-y, \quad x, 0), \text{ if } |z| < |x| \text{ and } |z| < |y| \end{cases} \\
\mathbf{v} &= \bar{\mathbf{v}}/\|\bar{\mathbf{v}}\| \\
\mathbf{w} &= \mathbf{u} \times \mathbf{v}.
\end{aligned}
$$

Then $\mathbf{u}, \mathbf{v}, \mathbf{w}$ is an orthonormal basis. As a simple example, consider $\mathbf{u} = (-2/7, 6/7, 3/7)$. In this case, $\bar{\mathbf{v}} = (0, -3/7, 6/7)$, $\mathbf{v} = \frac{1}{\sqrt{45}}(0, -3, 6)$, and $\mathbf{w} = \mathbf{u} \times \mathbf{v} = \frac{1}{7\sqrt{45}}(45, 12, 6)$.

3.1. Discussion

A more naive approach would be to simply compute $\mathbf{v} = \mathbf{e_1} \times \mathbf{u}$ and $\mathbf{u} = \mathbf{v} \times \mathbf{w}$. This becomes ill-behaved when $\mathbf{u}$ and $\mathbf{e_1}$ are nearly parallel, at which point the naive approach substitutes $\mathbf{e_2}$ for $\mathbf{e_1}$. One could also choose a random vector instead of $\mathbf{e_1}$, and this works with high probability. Our algorithm simply systematically avoids the problem with these two approaches.

Another naive approach is to apply the Gram-Schmidt process (see, for example, [Hoffman, Kunze 71], Section 8.2) to the set $\mathbf{u}, \mathbf{e_1}, \mathbf{e_2}, \mathbf{e_3}$ discarding any vector whose projection onto the subspace orthogonal to the prior ones is shorter than, say, 1/10th. This works too—in fact, it can be used for any number of dimensions—but uses multiple square roots, and hence is computationally expensive.

References

[Milnor 65] John Milnor. *Topology from the Differentiable Viewpoint*, Charlottesville: University Press of Virginia, 1965.

[Hoffman, Kunze 71] Kenneth Hoffman and Ray Kunze. *Linear Algebra*, Englewood Cliffs, NJ: Prentice Hall, 1971.

Web Information:

http://www.acm.org/jgt/papers/HughesMoller99

John F. Hughes, Brown University, Computer Science Department, 115 Waterman Street, Providence, RI 02912 (jfh@cs.brown.edu)

Tomas Möller, Chalmers University of Technology, Department of Computer Engineering, 412 96 Gothenburg, Sweden (tompa@acm.org) *Currently at U. C. Berkeley.*

Received December 10, 1999; accepted January 14, 2000.

Current Contact Information:

John F. Hughes, Brown University, Computer Science Department, 115 Waterman Street, Providence, RI 02912 (jfh@cs.brown.edu)

Tomas Akenine-Möller, Department of Computer Science, Lund University, Box 118, 221 00 Lund, Sweden (tam@cs.lth.se)

Vol. 4, No. 4: 1–4

Efficiently Building a Matrix to Rotate One Vector to Another

Tomas Möller

Chalmers University of Technology

John F. Hughes

Brown University

Abstract. We describe an efficient (no square roots or trigonometric functions) method to construct the 3×3 matrix that rotates a unit vector $\mathbf{f}$ into another unit vector $\mathbf{t}$, rotating about the axis $\mathbf{f} \times \mathbf{t}$. We give experimental results showing this method is faster than previously known methods. An implementation in C is provided.

1. Introduction

Often in graphics we have a unit vector, $\mathbf{f}$, that we wish to rotate to another unit vector, $\mathbf{t}$, by rotation in a plane containing both; in other words, we seek a rotation matrix $\mathbf{R}(\mathbf{f}, \mathbf{t})$ such that $\mathbf{R}(\mathbf{f}, \mathbf{t})\mathbf{f} = \mathbf{t}$. This paper describes a method to compute the matrix $\mathbf{R}(\mathbf{f}, \mathbf{t})$ from the coordinates of $\mathbf{f}$ and $\mathbf{t}$, without square root or trigonometric functions. Fast and robust C code can be found on the accompanying Web site.

2. Derivation

Rotation from $\mathbf{f}$ to $\mathbf{t}$ could be generated by letting $\mathbf{u} = \mathbf{f} \times \mathbf{t} / \|\mathbf{f} \times \mathbf{t}\|$, and then rotating about the unit vector $\mathbf{u}$ by $\theta = \arccos(\mathbf{f} \cdot \mathbf{t})$. A formula for the matrix that rotates about $\mathbf{u}$ by θ is given in Foley et al. [Foley et al. 90], namely

$$\begin{pmatrix} u_x^2 + (1 - u_x^2)\cos\theta & u_x u_y(1 - \cos\theta) - y_z\sin\theta & u_x u_y + u_y\sin\theta \\ u_x u_y(1 - \cos\theta) + u_x\sin\theta & u_y^2 + (1 - u_y^2)\cos\theta & u_y u_z(1 - \cos\theta) - u_x\sin\theta \\ u_x u_z(1 - \cos\theta) - u_y\sin\theta & u_y u_z(1 - \cos\theta) + u_x\sin\theta & u_z^2 + (1 - u_z^2)\cos\theta \end{pmatrix}$$

The above involves $\cos(\theta)$, which is just $\mathbf{f} \cdot \mathbf{t}$ and $\sin(\theta)$, which is $\|\mathbf{f} \times \mathbf{t}\|$. If we instead let

$$\mathbf{v} = \mathbf{f} \times \mathbf{t}$$
$$c = \mathbf{f} \cdot \mathbf{t}$$
$$h = \frac{1 - c}{1 - c^2} = \frac{1 - c}{\mathbf{v} \cdot \mathbf{v}} = \frac{1}{1 + c}$$

then, after considerable algebra, one can simplify the matrix to

$$\mathbf{R(f, t)} = \begin{pmatrix} c + hv_x^2 & hv_x v_y - v_z & hv_x v_z + v_y \\ hv_x v_y + v_z & c + hv_y^2 & hv_y v_z - v_x \\ hv_x v_z - v_y & hv_y v_z + v_x & c + hv_z^2 \end{pmatrix} \qquad (1)$$

Note that this formula for $\mathbf{R(f, t)}$ has no square roots or trigonometric functions.

When $\mathbf{f}$ and $\mathbf{t}$ are nearly parallel (i.e., if $|\mathbf{f} \cdot \mathbf{t}| > 0.99$), the computation of the plane that they define (and the normal to that plane, which will be the axis of rotation) is numerically unstable; this is reflected in our formula by the denominator of h becoming close to zero.

In this case, we observe that a product of two reflections (angle-preserving transformations of determinant -1) is always a rotation, and that reflection matrices are easy to construct: For any vector $\mathbf{u}$, the Householder matrix [Golub, Van Loan 96]

$$\mathbf{H(u)} = \mathbf{I} - \frac{2}{\mathbf{u} \cdot \mathbf{u}}\mathbf{u}\mathbf{u}^t$$

reflects the vector $\mathbf{u}$ to $-\mathbf{u}$, and leaves fixed all vectors orthogonal to $\mathbf{u}$. In particular, if $\mathbf{p}$ and $\mathbf{q}$ are unit vectors, then $\mathbf{H(q - p)}$ exchanges $\mathbf{p}$ and $\mathbf{q}$, leaving $\mathbf{p} + \mathbf{q}$ fixed.

With this in mind, we choose a unit vector $\mathbf{p}$ and build two reflection matrices: one that swaps $\mathbf{f}$ and $\mathbf{p}$, and the other that swaps $\mathbf{t}$ and $\mathbf{p}$. The product of these is a rotation that takes $\mathbf{f}$ to $\mathbf{t}$.

To choose $\mathbf{p}$, we determine which coordinate axis ($x, y,$ or z) is most nearly orthogonal to $\mathbf{f}$ (the one for which the corresponding coordinate of $\mathbf{f}$ is smallest in absolute value) and let $\mathbf{p}$ be a unit vector along that axis. We then build $\mathbf{A} = \mathbf{H(p - f)}$, and $\mathbf{B} = \mathbf{H(p - t)}$, and the rotation we want is $\mathbf{R} = \mathbf{BA}$.

That is, if we let

$$\mathbf{p} = \begin{cases} \hat{\mathbf{x}}, & \text{if } |f_x| < |f_y| \text{ and } |f_x| < |f_z| \\ \hat{\mathbf{y}}, & \text{if } |f_y| < |f_x| \text{ and } |f_y| < |f_z| \\ \hat{\mathbf{z}}, & \text{if } |f_z| < |f_x| \text{ and } |f_z| < |f_y| \end{cases}$$

$$\mathbf{u} = \mathbf{p} - \mathbf{f}$$

$$\mathbf{v} = \mathbf{p} - \mathbf{t}$$

then the entries of $\mathbf{R}$ are given by

$$r_{ij} = \delta_{ij} - \frac{2}{\mathbf{u} \cdot \mathbf{u}} u_i u_j - \frac{2}{\mathbf{v} \cdot \mathbf{v}} v_i v_j + \frac{4 \mathbf{u} \cdot \mathbf{v}}{(\mathbf{u} \cdot \mathbf{u})(\mathbf{v} \cdot \mathbf{v})} v_i u_j \qquad (2)$$

where $\delta_{ij} = 1$ when $i = j$ and $\delta_{ij} = 0$ when $i \neq j$.

3. Performance

We tested the new method for performance against all previously known (by the authors) methods for rotating a unit vector into another unit vector. A naive way to rotate $\mathbf{f}$ into $\mathbf{t}$ is to use quaternions to build the rotation directly: Letting $\mathbf{u} = \mathbf{v}/\|\mathbf{v}\|$, where $\mathbf{v} = \mathbf{f} \times \mathbf{t}$, and letting $\phi = (1/2) \arccos(\mathbf{f} \cdot \mathbf{t})$, we define $\mathbf{q} = (\sin(\phi)\mathbf{u}; cos\phi)$ and then convert the quaternion $\mathbf{q}$ into a rotation via the method described by Shoemake [Shoemake 85]. This rotation takes $\mathbf{f}$ to $\mathbf{t}$, and we refer to this method as Naive. The second is called Cunningham and is just a change of bases [Cunningham 90]. Goldman [Goldman 90] gives a routine for rotating around an arbitrary axis: in our third method we simplified his matrix for our purposes; this method is denoted Goldman. All three of these require that some vector be normalized; the quaternion method requires normalization of $\mathbf{v}$; the Cunningham method requires that one input be normalized, and then requires normalization of the cross-product. Goldman requires the normalized axis of rotation. Thus, the requirement of unit-vector input in our algorithm is not exceptional.

For the statistics below, we used 1,000 pairs of random normalized vectors $\mathbf{f}$ and $\mathbf{t}$. Each pair was fed to the matrix routines 10,000 times to produce accurate timings. Our timings were done on a Pentium II 400 MHz with compiler optimizations for speed on.

Routine:	Naive	Cunningham	Goldman	New Routine
Time (s):	18.6	13.2	6.5	4.1

The fastest of previous known methods (Goldman) still takes about 50% more time than our new routine, and the naive implementation takes almost 350%

more time. Similar performance can be expected on most other architectures, since square roots and trigonometric functions are expensive to use.

Acknowledgments. Thanks to Eric Haines for encouraging us to write this.

References

[Cunningham 90] Steve Cunningham. "3D Viewing arid Rotation using Orthonormal Bases." In *Graphics Gems*, edited by Andrew S. Glassner, pp. 516–521, Boston: Academic Press, Inc., 1990.

[Foley et al. 90] J. D. Foley, A. van Dam, S. K. Feiner, and J. F. Hughes. *Computer Graphics—Principles and Practice,* Second Edition, Reading, MA: Addison-Wesley, 1990.

[Goldman 90] Ronald Goldman. "Matrices and Transformations." In *Graphics Gems,* edited by Andrew S. Glassner, pp. 472-475, Boston: Academic Press, Inc., 1990.

[Golub, Van Loan 96] Gene Golub and Charles Van Loan. *Matrix Computations,* Third Edition, Baltimore: Johns Hopkins University Press, 1996.

[Shoemake 85] Ken Shoemake. "Animating Rotation with Quaternion Curves." *Computer Graphics (Proc. SIGGRAPH '85),* 19(3): 245–254 (July 1985).

Web Information:

http://www.acm.org/jgt/papers/MollerHughes99

Tomas Möller, Chalmers University of Technology, Department of Computer Engineering, 412 96 Gothenburg, Sweden (tompa@acm.org) *Currently at U. C. Berkeley.*

John F. Hughes, Brown University, Computer Science Department, 115 Waterman Street, Providence, RI 02912 (jfh@cs.brown.edu)

Received June 28, 1999; accepted in revised form January 14, 2000.

Current Contact Information:

Tomas Akenine-Möller, Department of Computer Science, Lund University, Box 118, 221 00 Lund, Sweden (tam@cs.lth.se)

John F. Hughes, Brown University, Computer Science Department, 115 Waterman Street, Providence, RI 02912 (jfh@cs.brown.edu)

Vol. 4, No. 2: 37–43

Fast Projected Area Computation for Three-Dimensional Bounding Boxes

Dieter Schmalstieg and Robert F. Tobler
Vienna University of Technology

Abstract. The area covered by a three-dimensional bounding box after projection onto the screen is relevant for view-dependent algorithms in real-time and photorealistic rendering. We describe a fast method to compute the accurate two-dimensional area of a three-dimensional oriented bounding box, and show how it can be computed equally fast or faster than its approximation with a two-dimensional bounding box enclosing the projected three-dimensional bounding box.

1. Introduction

Computer graphics algorithms using heuristics, like level of detail (LOD) selection algorithms, make it sometimes necessary to estimate the area an object covers on the screen after perspective projection [Funkhouser, Sequin 93]. Doing this exactly would require first drawing the object and then counting the covered pixels, which is quite infeasible for real-time applications. Instead, oftentimes a bounding box (bbox) is used as a rough estimate: The bbox of the object is projected to the screen, and its size is taken.

Another application area is view-dependent hierarchical radiosity, where a fast method for calculating the projection area can be used to estimate the importance of high level patches, obviating the need to descend the hierarchy in places of little importance.

The reason for favoring bounding boxes over bounding spheres is that they provide a potentially tighter fit (and hence a better approximation) for the object while offering roughly the same geometric complexity as spheres. How-

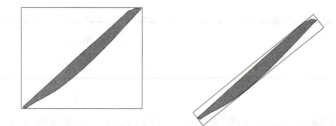

Figure 1. Axis-aligned bounding boxes (left) are often inferior to oriented bounding boxes (right).

ever, usually axis-aligned bboxes are used, which can also be a poor fit for the enclosed object. In contrast, an oriented bounding box (OBB) requires an additional transformation to be applied, but allows a comparatively tight fit (Figure 1). The speed and applicability of OBBs in other areas has been shown by Gottschalk et al. [Gottschalk et al. 96]. For the construction of an OBB, refer to [Wu 92]. To estimate the two-dimensional area of a three-dimensional object when projected to the screen, one can find the perspective projection of the corners of an axis-aligned bbox, and then use the area of the rectangle (two-dimensional bbox) enclosing the three-dimensional bbox to estimate the area of the object on the screen. This procedure entails two nested approximations which are not necessarily a tight fit, and the error can be large.

Instead, we directly project an OBB and compute the area of the enclosing two dimensional polygon. This procedure yields significantly better approximations. Moreover, we will show that the procedure can be coded to require fewer operations than the nested bbox approach.

2. Algorithm Overview

In this section, we will show how a simple viewpoint classification leads to an approach driven by a lookup table, followed by an area computation based on a contour integral. Both steps can be coded with few operations and are computationally inexpensive. When a three-dimensional box is projected to the screen either one, two, or three adjacent faces are visible, depending on the viewpoint (Figure 2):

- **Case 1:** one face visible, two-dimensional hull polygon consists of four vertices

- **Case 2:** two faces visible, two-dimensional hull polygon consists of six vertices

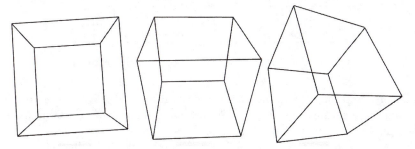

Figure 2. One, two, or three faces of a box may be visible.

- **Case 3:** three faces visible, two-dimensional hull polygon consists of six vertices

Whether a particular placement of the viewpoint relative to the bbox yields Case 1, 2, or 3, can be determined by examining the position of the viewpoint with respect to the six planes defined by the six faces of the bbox. These six planes subdivide Euclidean space into $3^3 = 27$ regions. The case where the viewpoint is inside the box does not allow meaningful area computation, so 26 valid cases remain.

By classifying the viewpoint as left or right of each of the six planes, we obtain $2^6 = 64$ theoretical cases, of which 26 are valid. For each of these cases we describe the hull polygon as an ordered set of vertex indices which can be precomputed and stored in a two-dimensional lookup table, called the hull vertex table.

An efficient implementation of the classification is to transform the viewpoint in the local coordinate system of the OBB, where each of the planes is parallel to one of the major planes, and the classification can be made by comparing one scalar value.

After the classification, the area of the hull polygon must be computed from the bbox vertices given in the hull vertex table. Our sample implementation uses a fast contour integral [Foley et al. 90].

3.   Implementation

For an efficient implementation, the central data structure is the hull vertex table. It stores the ordered vertices that form the outline of the hull polygon after projection to two dimensions, as well as the number of vertices in the outline (four or six, with zero indicating an invalid case). The table is indexed with a 6-bit code according to Table 1.

Bit	5	4	3	2	1	0
Code	back	front	top	bottom	right	left

Table 1. Bit code used to index into the hull vertex table.

By precomputing this table, many computational steps can be saved when a bounding box area is computed at runtime. The hull vertex table used in the sample implementation is shown in Table 2.

Using this hull vertex table (`hullvertex`), the following C function `calculateBoxArea` computes the projected area of an OBB from the viewpoint given in parameter `eye` and the bounding box `bbox` given as an array of eight vertices (both given in local bbox coordinates). We assume an auxiliary function `projectToScreen`, which performs perspective projection of an OBB vertex to screen space.

```
float calculateBoxArea(Vector3D eye, Vector3D bbox[8])
{
    Vector2D dst[8]; float sum; int pos, num, i;
    int pos = ((eye.x < bbox[0].x)      )    // 1 = left    |  compute 6-bit
            + ((eye.x > bbox[7].x) << 1)    // 2 = right   |       code to
            + ((eye.y < bbox[0].y) << 2)    // 4 = bottom  |   classify eye
            + ((eye.y > bbox[7].y) << 3)    // 8 = top     |with respect to
            + ((eye.z < bbox[0].z) << 4)    // 16 = front  | the 6 defining
            + ((eye.z > bbox[7].z) << 5);   // 32 = back   |       planes
    if (!num = hullvertex[pos][6]) return -1.0;   //look up number of vertices
                                                  //return -1 if inside
    for(i=0; i<num; i++) dst[i]:= projectToScreen(bbox[hullvertex[pos][i]]);
    sum = (dst[num-1].x - dst[0].x) * (dst[num-1].y + dst[0].y);
    for (i=0; i<num-1; i++)
        sum += (dst[i].x - dst[i+1].x) * (dst[i].y + dst[i+1].y);
    return sum * 0.5;                             //return corrected value
}
```

4.   Discussion

The proposed implementation gives superior results to a simple "two-dimensional bbox of three-dimensional bbox" implementation. However, although it yields better accuracy, it can be implemented to use slightly fewer operations than the simple two-dimensional bbox variant. Our algorithm is composed of the following steps:

1. Transformation of the viewpoint into local bbox coordinates: Note that this step is not included in the sample implementation. Given the transformation matrix from world coordinates to local bbox coordinates, this is a simple affine transformation a three-dimensional

Case	Num	Vertex Indices						Description
0	0							inside
1	4	0	4	7	3			left
2	4	1	2	6	5			right
3	0							-
4	4	0	1	5	4			bottom
5	6	0	1	2	6	5	4	bottom, right
6	6	0	1	2	6	5	4	bottom, right
7	0							-
8	4	2	3	7	6			top
9	6	4	7	6	2	3	0	top, left
10	6	2	3	7	6	5	1	top, right
11	0							-
12	0							-
13	0							-
14	0							-
15	0							-
16	4	0	3	2	1			front
17	6	0	4	7	3	2	1	front, left
18	6	0	3	2	6	5	1	front, right
19	0							-
20	6	0	3	2	1	5	4	front, bottom
21	6	1	5	4	7	3	2	front, bottom, left
22	6	0	3	2	6	5	4	front, bottom, right
23	0							-
24	6	0	3	7	6	2	1	front, top
25	6	0	4	7	6	2	1	front, top, left
26	6	0	3	7	6	5	1	front, top, right
27	0							-
28	0							-
29	0							-
30	0							-
31	0							-
32	4	4	5	6	7			back
33	6	4	5	6	7	3	0	back, left
34	6	1	2	6	7	4	5	back, right
35	0							-
36	6	0	1	5	6	7	4	back, bottom
37	6	0	1	5	6	7	3	back, bottom, left
38	6	0	1	2	6	7	4	back, bottom, right
39	0							-
40	6	2	3	7	4	5	6	back, top
41	6	0	4	5	6	2	3	back, top, left
42	6	1	2	3	7	4	5	back, top, right
≥ 43	0							-

Table 2. The hull vertex table stores precomputed information about the projected bbox.

vector is multiplied with a 3×4 matrix, using 12 multiplications and nine additions.

2. Computation of the index into the hull vertex table: To perform this step, the viewpoint's coordinates are compared to the defining planes of the bounding box. As there are six planes, this step uses at most six comparisons (a minimum of three comparisons is necessary if a cascading conditional is used for the implementation).

3. Perspective projection of the hull vertices: This step has variable costs depending on whether the hull consists of four or six vertices. The three-dimensional vertices that form the hull polygon must be projected into screen space. This is a perspective projection, but the fact that we are only interested in the x and y components (for area computation) allows a few optimizations. The x and y components are transformed using three multiply and two add operations per component. However, a perspective projection requires normalization after the matrix multiplication to yield homogeneous coordinates. A normalization factor must be computed, which takes one perspective division and one add operation. The x and y components are then normalized, taking two multiply operations. This analysis yields a total of 18 operations for an optimized perspective projection.

4. Area computation using a contour integral: Each signed area segment associated with one edge requires one add, one subtract, and one multiply operation, plus one add operation for the running score, except for the first edge. The result must be divided by two (one multiply operation). The total number of operations again depends on whether there are four or six vertices (and edges).

The total number of operations is 159 for Case 2 and 3 (six vertices), and 115 for case 1 (four vertices). The number of operations required to compute a simple two-dimensional box area is 163. It can be computed as follows: Projection of eight vertices (18×8 operations), computation of the two-dimensional bbox of the projected vertices using min-max tests (2×8 comparisons), two-dimensional box computation (two subtract, one add, one multiply operation).

As the number of operations required to compute the exact projected area of a three-dimensional bbox is of the same order or even less expensive than the simple approach using a two-dimensional bbox, it is recommended to use this procedure for real-time bounding box area computation.

Acknowledgments. Special thanks to Erik Pojar for his help with the sample implementation. This work was sponsored by the Austrian Science Foundation (FWF) under contract no. P-11392-MAT.

References

[Foley et al. 90] J. Foley, A. van Dam, S. Feiner, and J. Hughes. *Computer Graphics—Principles and Practice.* 2nd edition, Reading, MA: Addison Wesley, 1990.

[Funkhouser, Sequin 93] T. A. Funkhouser and C. H. Sequin. "Adaptive Display Algorithm for Interactive Frame Rates During Visualisation of Complex Virtual Environments." In *Proceedings of SIGGRAPH 93, Computer Graphics Proceedings, Annual Conference Series,* edited by James T. Kajiya, pp. 247–254, New York: ACM Press, 1993.

[Gottschalk et al. 96] S. Gottschalk, M. Lin, and D. Manchoa. "OBBTree: A Hierarchical Structure for Rapid Interference Detection." In *Proceedings of SIGGRAPH 96, Computer Graphics Proceedings, Annual Conference Series,* edited by Holly Rushmeier, pp. 171–180, Reading, MA: Addison-Wesley, 1996.

[Wu 92] X. Wu: "A Linear-time Simple Bounding Volume Algorithm." In *Graphics Gems II,* edited by David Kirk, pp. 301–306, Boston: Academic Press, 1992.

Web Information:

The C source code in this paper, including the hull vertex table data, is available online at http://www.acm.org/jgt/papers/SchmalstiegTobler99

Dieter Schmalstieg, Vienna University of Technology, Karlsplatz 13/186/2, A-1040 Vienna, Austria (dieter@cg.tuvien.ac.at)

Robert F. Tobler, Vienna University of Technology, Karlsplatz 13/186/2, A-1040 Vienna, Austria (rft@cg.tuvien.ac.at)

Received February 19, 1999; accepted August 27, 1999

Current Contact Information:

Dieter Schmalstieg, Graz University of Technology, Inffeldgasse 16, A-8010 Graz, Austria (schmalstieg@icg.tu-graz.ac.at)

Robert F. Tobler, VRVis Research Center, Donau-City-Strasse 1/3 OG, A-1030 Vienna, Austria (tobler@vrvis.at)

Vol. 8, No. 1: 3–15

Computing a View Frustum to Maximize an Object's Image Area

Kok-Lim Low and Adrian Ilie
University of North Carolina at Chapel Hill

Abstract. This paper presents a method to compute a view frustum for a three-dimensional object viewed from a given viewpoint, such that the object is completely enclosed in the frustum and the object's image area is also near-maximal in the given two-dimensional rectangular viewing region. This optimization can be used to improve the resolution of shadow and texture maps for projective texture mapping. Instead of doing the optimization in three-dimensional space to find a good view frustum, our method uses a two-dimensional approach. The basic idea of our approach is as follows. First, from the given viewpoint, a conveniently computed view frustum is used to project the three-dimensional vertices of the object to their corresponding two-dimensional image points. A tight two-dimensional bounding quadrilateral is then computed to enclose these two-dimensional image points. Next, considering the projective warp between the bounding quadrilateral and the rectangular viewing region, our method applies a technique of camera calibration to compute a new view frustum that generates an image that covers the viewing region as much as possible.

1. Introduction

In interactive computer graphics rendering, we often need to compute a view frustum from a given viewpoint such that a selected three-dimensional object or a group of three-dimensional objects is totally inside the rendered two-dimensional rectangular image. This kind of view frustum computation is usually needed when generating shadow maps [Williams 78] from light sources, and images for projective texture mapping [Segal et al. 92], [Hoff 98].

31

Vol. 8, No. 1: 3–15

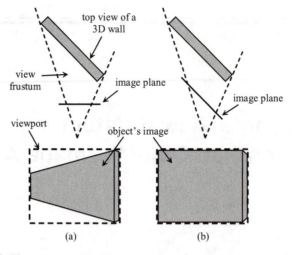

Figure 1. (a) The symmetric perspective view frustum cannot enclose the three-dimensional object tightly enough, therefore, the object's image does not utilize efficiently the viewport area. (b) By manipulating the view frustum such that the image plane becomes parallel to the larger face of the three-dimensional wall, we can improve the object's image to cover almost the whole viewport

The easiest way to compute such a view frustum is to precompute a simple three-dimensional bounding volume, such as a bounding sphere, around the three-dimensional object, and create a *symmetric* perspective view frustum that encloses the object's bounding volume. However, very often, this view frustum is not enclosing the three-dimensional object tightly enough to produce an image of the object that covers the two-dimensional rectangular viewing region as much as possible. We will refer to the two-dimensional rectangular viewing region as the *viewport*, and the projection of the three-dimensional object in the viewport as the *object's image*. If the object's image is too small, we are not efficiently utilizing the available viewport area to produce a shadow map or projective texture map that could have higher resolution due to a larger image of the object. A small image region of the object in a shadow map usually results in blocky shadow edges, and similarly, a low-resolution image region in a texture map can also result in a blocky rendered image.

Other methods increase the object's image area in the viewport by using a tighter three-dimensional bounding volume, such as the three-dimensional convex hull of the object [Berg et al. 00]. However, this is computationally expensive, and there is still a lot of room for improvement by manipulating the shape of the view frustum and the orientation of the image plane. Figure 1 shows an example.

This paper presents a method to compute a view frustum for a three-dimensional object viewed from a given viewpoint, such that the object's image is entirely inside the viewport and its area is also near-maximal. For computational efficiency, our method does not seek to compute the optimal view frustum, but to compromise for one that is near-optimal. Instead of doing the optimization in three-dimensional space to find a good view frustum, our method uses a two-dimensional approach. This makes the method more efficient and simpler to implement.

2. Overview of the Method

Without loss of generality, we will describe our method in the context of the OpenGL API [Woo et al. 99]. In OpenGL, defining a view frustum from an arbitrary viewpoint requires the definition of two transformations: the *view transformation*, which transforms points in the world coordinate system into the eye coordinate system; and the *projection transformation*, which transforms points in the eye coordinate system into the normalized device coordinate (NDC) system.

Given a viewpoint, a three-dimensional object in the world coordinate system, and the viewport's width and height, our objective is to compute a valid view frustum (i.e., a view transformation and a projection transformation) that maximizes the area of the object's image in the viewport. We provide an overview of our method below and the key steps are described in more detail in Sections 3 and 4.

2.1. Start with an Initial Frustum

We start with a conveniently computed view frustum by bounding the object with a sphere and then creating a symmetric perspective view frustum that encloses the sphere. The view transformation and the projection transformation that represent this frustum can be readily obtained using the OpenGL commands glGetDoublev(GL_MODELVIEW_MATRIX, m) and glGetDoublev(GL_PROJECTION_MATRIX, p), respectively.

We use the two transformations and the viewport settings to explicitly transform all the three-dimensional vertices of the object from the world coordinates to their corresponding two-dimensional image points.

2.2. Compute a Tight Bounding Quadrilateral (Section 3)

We compute a tight bounding quadrilateral of the two-dimensional image points. The basic idea, illustrated in Figure 2, is to start by computing a two-dimensional convex hull of the image points, and then incrementally removing the edges of the convex hull until a bounding quadrilateral remains.

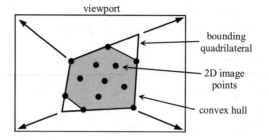

Figure 2. The three-dimensional vertices of the object are first projected onto their corresponding two-dimensional image points. A two-dimensional convex hull is computed for these image points, and it is then incrementally reduced to a quadrilateral. The bounding quadrilateral is related to the viewport's rectangle by a projective warp. This warping effect can be achieved by rotating and moving the image plane.

2.3. *Compute an Optimized View Frustum (Section 4)*

The most important idea of our method lies in the observation that the bounding quadrilateral and the rectangular viewport are related only by a projective warp or two-dimensional collineation (see Chapter 2 of [Faugeras 93]). Equally important to know is that this projective warp from the bounding quadrilateral to a rectangle can be achieved by merely rotating and moving the image plane.

3. Computing a Tight Bounding Quadrilateral

We start by computing the two-dimensional convex hull of the two-dimensional image points, using methods such as Graham's algorithm [O'Rourke 98]. The time complexity of this step is $O(m \log m)$, where m is the number of image points.

Aggarwal et al. presented a general technique to compute the smallest convex k-sided polygon to enclose a given convex n-sided polygon [Aggarwal et al. 98]. Their method runs in $O(n^2 \log n \log k)$ time, however, and can be difficult to implement.

Here, we describe an alternative algorithm to compute a convex bounding quadrilateral. Our algorithm produces only near-optimal results, but is simple to implement and has time complexity $O(n \log n)$.

Our algorithm obtains the convex bounding quadrilateral by iteratively removing edges from the convex hull using a greedy approach until only four edges remain. To remove an edge i, we need to first make sure that the sum of the interior angles it makes with the two adjacent edges is more than $180°$.

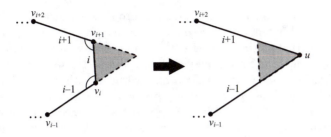

Figure 3. Removing edge i.

Then, we extend the two adjacent edges toward each other to intersect at a point (see Figure 3).

For a convex hull with n edges, we iterate $n - 4$ times, removing one edge each time to yield a quadrilateral. At each iteration, we choose to remove the edge whose removal would add the smallest area to the resulting polygon. For example, in Figure 3, removing edge i adds the gray-shaded area to the resulting polygon. We use a heap to find the edge to remove in constant time. After the edge is removed, we must update the area values of its two adjacent edges. Since initially building the heap requires $O(n \log n)$ time, and each iteration has two $O(\log n)$ heap updates, the time complexity of our algorithm is $O(n \log n)$.

It can be easily proved that for any convex polygon of five or more sides, there always exists at least one edge that can be removed. Since the resulting polygon is also a convex polygon, by induction, we can always reduce the initial input convex hull to a convex quadrilateral.

Of course, if the initial convex hull is already a quadrilateral, we do not need to do anything. If the initial convex hull is a triangle, we just create a bounding parallelogram whose diagonal corresponds to the longest edge of the triangle, and three of its corners coincide with the three corners of the triangle. This ensures that the object's image occupies half the viewport, which is the optimal area in this case.

4. Computing an Optimized View Frustum

After we have found a tight bounding quadrilateral, we want to compute a view frustum that warps the quadrilateral to the viewport's rectangle as illustrated in Figure 2.

First, we need to decide to which corner of the viewport's rectangle each quadrilateral corner is to be warped. We have chosen to match the longest edge and its opposite edge of the quadrilateral with the longer edges of the viewport's rectangle.

Using the view transformation and the projection transformation of the conveniently computed view frustum, we inverse-project each corner of the bounding quadrilateral back into the three-dimensional world coordinate system as a ray originating from the viewpoint. Taking the world coordinates of any three-dimensional point on the ray and pairing it with the two-dimensional *pixel coordinates* of the corresponding corner of the viewport's rectangle, we get a *pair-correspondence*. With four pair-correspondences, one for each corner, we are able to use a camera calibration technique to solve for the desired view frustum.

4.1. A Camera Calibration Technique

For a pinhole camera, which is the camera model used in OpenGL, the effect of transforming a three-dimensional point in the world coordinate system into a two-dimensional image point in the viewport can be described by the following expression:

$$
\begin{pmatrix} u_i \\ v_i \\ w_i \end{pmatrix} = \mathbf{P} \cdot \begin{pmatrix} X_i \\ Y_i \\ Z_i \\ 1 \end{pmatrix} = \begin{pmatrix} a & 0 & c_x \\ 0 & b & c_y \\ 0 & 0 & 1 \end{pmatrix} \cdot \begin{pmatrix} r_{11} & r_{12} & r_{13} & t_1 \\ r_{21} & r_{22} & r_{23} & t_2 \\ r_{31} & r_{32} & r_{33} & t_3 \end{pmatrix} \cdot \begin{pmatrix} X_i \\ Y_i \\ Z_i \\ 1 \end{pmatrix}
\tag{1}
$$

where

- a, b, c_x, and c_y are collectively called the *intrinsic parameters* of the camera,

- r_{ij} and t_i, respectively, define the *rotation* and *translation* of the view transformation, and they are called the *extrinsic parameters* of the camera,

- $(X_i, Y_i, Z_i, 1)^T$ are the homogeneous coordinates of a point in the world coordinate system, and

- the pixel coordinates of the two-dimensional image point are

$$
\begin{pmatrix} x_i \\ y_i \end{pmatrix} = \begin{pmatrix} u_i/w_i \\ v_i/w_i \end{pmatrix}.
\tag{2}
$$

$\mathbf{P}$ is a 3×4 projection matrix. Note that this is not the same as OpenGL's projection transformation: $\mathbf{P}$ maps a three-dimensional point in the world coordinate system to two-dimensional pixel coordinates, whereas OpenGL's projection transformation maps a three-dimensional point in the eye coordinate system to a three-dimensional point in the NDC. (Later on, we will describe how to construct OpenGL's projection matrix from $\mathbf{P}$.)

Since the viewpoint's position is known, we can first apply a translation to the world coordinate system such that the viewpoint is now located at the origin. We will refer to this as the *shifted world coordinate system*, and with it, we can simplify (1) to

$$
\begin{pmatrix} u_i \\ v_i \\ w_i \end{pmatrix} = \mathbf{P} \cdot \begin{pmatrix} X_i \\ Y_i \\ Z_i \end{pmatrix} = \begin{pmatrix} a & 0 & c_x \\ 0 & b & c_y \\ 0 & 0 & 1 \end{pmatrix} \cdot \begin{pmatrix} r_{11} & r_{12} & r_{13} \\ r_{21} & r_{22} & r_{23} \\ r_{31} & r_{32} & r_{33} \end{pmatrix} \cdot \begin{pmatrix} X_i \\ Y_i \\ Z_i \end{pmatrix} \quad (3)
$$

where $\mathbf{P}$ is now a 3×3 matrix, and $(X_i, Y_i, Z_i)^T$ are the three-dimensional coordinates of a point in the shifted world coordinate system.

To solve for the intrinsic and extrinsic camera parameters, we will first solve for the matrix $\mathbf{P}$, and then decompose $\mathbf{P}$ into the individual camera parameters.

4.1.1. Solving for the Projection Matrix

If we write $\mathbf{P}$ as

$$
\mathbf{P} = \begin{pmatrix} p_{11} & p_{12} & p_{13} \\ p_{21} & p_{22} & p_{23} \\ p_{31} & p_{32} & p_{33} \end{pmatrix}, \quad (4)
$$

then the pixel coordinates of the ith two-dimensional image point can be written as

$$
x_i = \frac{u_i}{w_i} = \frac{p_{11}X_i + p_{12}Y_i + p_{13}Z_i}{p_{31}X_i + p_{32}Y_i + p_{33}Z_i}
$$

$$
y_i = \frac{v_i}{w_i} = \frac{p_{21}X_i + p_{22}Y_i + p_{23}Z_i}{p_{31}X_i + p_{32}Y_i + p_{33}Z_i}. \quad (5)
$$

We can rearrange (5) to get

$$
p_{11}X_i + p_{12}Y_i + p_{13}Z_i - x_i(p_{31}X_i + p_{32}Y_i + p_{33}Z_i) = 0
$$

$$
p_{21}X_i + p_{22}Y_i + p_{23}Z_i - y_i(p_{31}X_i + p_{32}Y_i + p_{33}Z_i) = 0. \quad (6)
$$

Because of the divisions u_i/w_i and v_i/w_i in (5), $\mathbf{P}$ is defined up to an arbitrary scale factor, and has only eight independent entries. Therefore, the four pair-correspondences we have previously obtained are sufficient to solve for $\mathbf{P}$. Note that because of the removal of the translation in (3), the three-dimensional point in each pair-correspondence must now be translated into the shifted world coordinate system. Note that by construction, our bounding quadrilateral is strictly convex, so no three corners will be collinear and we do not need to worry about degeneracy.

With the four pair-correspondences, we can form a homogeneous linear system

$$\mathbf{A} \cdot \mathbf{p} = \mathbf{0}, \tag{7}$$

where

$$\mathbf{p} = (p_{11}, p_{12}, p_{13}, p_{21}, p_{22}, p_{23}, p_{31}, p_{32}, p_{33})^{\mathrm{T}} \tag{8}$$

and

$$\mathbf{A} = \begin{pmatrix}
X_1 & Y_1 & Z_1 & 0 & 0 & 0 & -x_1 X_1 & -x_1 Y_1 & -x_1 Z_1 \\
0 & 0 & 0 & X_1 & Y_1 & Z_1 & -y_1 X_1 & -y_1 Y_1 & -y_1 Z_1 \\
X_2 & Y_2 & Z_2 & 0 & 0 & 0 & -x_2 X_2 & -x_2 Y_2 & -x_2 Z_2 \\
0 & 0 & 0 & X_2 & Y_2 & Z_2 & -y_2 X_2 & -y_2 Y_2 & -y_2 Z_2 \\
X_3 & Y_3 & Z_3 & 0 & 0 & 0 & -x_3 X_3 & -x_3 Y_3 & -x_3 Z_3 \\
0 & 0 & 0 & X_3 & Y_3 & Z_3 & -y_3 X_3 & -y_3 Y_3 & -y_3 Z_3 \\
X_4 & Y_4 & Z_4 & 0 & 0 & 0 & -x_4 X_4 & -x_4 Y_4 & -x_4 Z_4 \\
0 & 0 & 0 & X_4 & Y_4 & Z_4 & -y_4 X_4 & -y_4 Y_4 & -y_4 Z_4
\end{pmatrix}. \tag{9}$$

For the homogeneous system $\mathbf{A} \cdot \mathbf{p} = \mathbf{0}$, the vector $\mathbf{p}$ can be computed using SVD (singular value decomposition) related techniques as the eigenvector corresponding to the only zero eigenvalue of $\mathbf{A}^{\mathrm{T}}\mathbf{A}$. In other words, if the SVD of $\mathbf{A}$ is $\mathbf{U}\mathbf{D}\mathbf{V}^{\mathrm{T}}$, then $\mathbf{p}$ is the column of $\mathbf{V}$ corresponding to the only zero singular value of $\mathbf{A}$. For more details about camera calibration, see [Trucco, Verri 98], and for a comprehensive introduction to linear algebra and SVD, see [Strang 88]. An implementation of SVD can be found in [Press et al. 93].

4.1.2. Computing Camera Parameters

From the computed projection matrix $\mathbf{P}$, we want to express the intrinsic and extrinsic parameters as closed-form functions of the matrix entries. Recall that the computed matrix is defined only up to an arbitrary scale factor, therefore, to use the relationship

$$\mathbf{P} = \begin{pmatrix}
a r_{11} + c_x r_{31} & a r_{12} + c_x r_{32} & a r_{13} + c_x r_{33} \\
b r_{21} + c_y r_{31} & b r_{22} + c_y r_{32} & b r_{23} + c_y r_{33} \\
r_{31} & r_{32} & r_{33}
\end{pmatrix}, \tag{10}$$

we must first properly normalize $\mathbf{P}$. We observe that the last row in the matrix above corresponds to the last row of the rotation matrix, which must be of unit length. So we normalize $\mathbf{P}$ by dividing it by $\pm\sqrt{p_{31}^2 + p_{32}^2 + p_{33}^2}$, with the choice of sign still arbitrary.

We can now extract the camera parameters. For clarity, we write the three rows of $\mathbf{P}$ as the following column vectors:

$$\begin{aligned}
\mathbf{p}_1 &= (p_{11}, p_{12}, p_{13})^{\mathrm{T}}, \\
\mathbf{p}_2 &= (p_{21}, p_{22}, p_{23})^{\mathrm{T}}, \\
\mathbf{p}_3 &= (p_{31}, p_{32}, p_{33})^{\mathrm{T}}.
\end{aligned} \tag{11}$$

The values of the parameters can be computed as follows:

$$
\begin{aligned}
c_x &= \mathbf{p}_1^{\mathrm{T}} \mathbf{p}_3, \\
c_y &= \mathbf{p}_2^{\mathrm{T}} \mathbf{p}_3, \\
a &= -\sqrt{\mathbf{p}_1^{\mathrm{T}} \mathbf{p}_1 - c_x^2}, \\
b &= -\sqrt{\mathbf{p}_2^{\mathrm{T}} \mathbf{p}_2 - c_y^2}, \\
(r_{11}, r_{12}, r_{13})^{\mathrm{T}} &= \frac{(\mathbf{p}_1 - c_x \mathbf{p}_3)}{a}, \\
(r_{21}, r_{22}, r_{23})^{\mathrm{T}} &= \frac{(\mathbf{p}_2 - c_y \mathbf{p}_3)}{b} \\
(r_{31}, r_{32}, r_{33})^{\mathrm{T}} &= \mathbf{p}_3.
\end{aligned}
\tag{12}
$$

The sign λ of the normalization affects only the values of r_{ij}. It can be determined as follows. First, we use the rotation matrix $[r_{ij}]$ computed in the above procedure to transform the four shifted world points in the pair-correspondences. Since these three-dimensional points are all in front of the camera, their transformed z-coordinates should be negative, because the camera is looking in the $-z$ direction in the eye coordinate system. If it is not the case, we correct the r_{ij} by changing their signs.

4.1.3. Conversion to OpenGL Matrices

From the camera parameters obtained above, the OpenGL view transformation matrix is

$$
\mathbf{M}_{\mathrm{MODELVIEW}} = \begin{pmatrix} r_{11} & r_{12} & r_{13} & (-r_{11}v_x - r_{12}v_y - r_{13}v_z) \\ r_{21} & r_{22} & r_{23} & (-r_{21}v_x - r_{22}v_y - r_{23}v_z) \\ r_{31} & r_{32} & r_{33} & (-r_{31}v_x - r_{32}v_y - r_{33}v_z) \\ 0 & 0 & 0 & 1 \end{pmatrix},
\tag{13}
$$

where $(v_x, v_y, v_z)^{\mathrm{T}}$ is the position of the viewpoint in the world coordinate system.

The OpenGL projection matrix is

$$
\mathbf{M}_{\mathrm{PROJECTION}} = \begin{pmatrix} \frac{-2a}{W} & 0 & 1 - \frac{2c_x}{W} & 0 \\ 0 & \frac{-2b}{H} & 1 - \frac{2c_y}{H} & 0 \\ 0 & 0 & \frac{-(f+n)}{f-n} & \frac{-2fn}{f-n} \\ 0 & 0 & -1 & 0 \end{pmatrix},
\tag{14}
$$

where W and H are the width and height of the viewport in pixels, respectively, and n and f are the distances of the near and far plane from the viewpoint, respectively. If n and f cannot be known beforehand, a simple and efficient way to compute good values for n and f is to transform the bounding sphere of the three-dimensional object into the eye coordinate system and compute

$$
\begin{aligned}
n &= -o_z - r, \\
f &= -o_z + r,
\end{aligned}
\tag{15}
$$

where o_z is the z-coordinate of the center of the sphere in the eye coordinate system, and r is the radius of the sphere.

5.   Examples

In Figure 4, we show three example results. The images in the left-most column were generated using symmetric perspective view frusta enclosing the bounding spheres of the respective objects. The middle column shows the bounding quadrilaterals computed using our algorithm described in Section 3. The right-most column shows the images generated using the new frusta computed using our method. Note that each object is always viewed from the same viewpoint for both the unoptimized and optimized view frusta.

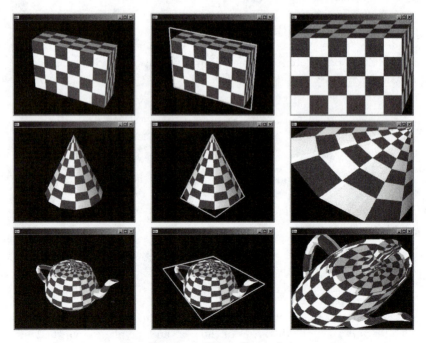

Figure 4. Example results.

6. Discussion

If the viewpoint is dynamic, a new view frustum has to be computed for every rendered frame. In the computation of the two-dimensional convex hull and the bounding quadrilateral, if the number of two-dimensional image points is too large, it may be difficult to render at interactive rates. For a nondeformable three-dimensional object, we can first precompute its three-dimensional convex hull, and project only the three-dimensional vertices of the convex hull onto the viewport as two-dimensional image points. This will reduce the number of two-dimensional points with which our algorithm needs to work. If the three-dimensional convex hull is still too complex, we can simplify it by reducing its number of faces and vertices. Note that the simplified hull should totally contain the original convex hull. The three-dimensional convex hull and its simplified version would be computed in a preprocessing step.

Besides the advantage of increasing the resolution of the object's image, our method can also improve the temporal consistency of the object's image resolution from frame to frame. If the three-dimensional object has a predominantly large face (or a predominant silhouette), the image plane of the computed view frustum will tend to be oriented with it for many viewpoints. This results in a more stable image plane, and therefore, more consistent object's image resolution. This benefit is important to projector-based displays in which projective texture mapping is used to produce perspective-correct imagery for the tracked users [Raskar et al. 98]. In this application, texture maps are generated from the user's viewpoint, and are then texture-mapped onto the display surfaces using projective texture mapping. Excessive changes in texture map resolution when the viewpoint moves can cause undesired effects in the projected imagery.

Something we wish we had done is to prove how much worse our approximated smallest enclosing quadrilaterals are, compared to the truly optimal ones. Such a proof would most likely be nontrivial. Since we also did not have an implementation of the algorithm described in [Aggarwal et al. 98] available to us, we could not do any empirical comparisons between our approximations and the true minimum areas. However, from manual inspection of our results, our algorithm always produced results that are within our expectation of being good approximations of the smallest possible quadrilaterals. Note that even if the quadrilateral is the smallest possible, it still cannot guarantee that the object's image area will be the largest possible. This is because the projective warp does not "scale" every part of the quadrilateral uniformly.

Raskar described a method to append a matrix that represents a two-dimensional collineation to an OpenGL projection matrix to achieve the desired projective warp of the original image [Raskar 99]. Though such a two-dimensional projective warp preserves collinearity in the two-dimensional image plane, it does not preserve collinearity in the three-dimensional NDC. This results in incorrect depth interpolation, and therefore, incorrect interpolation of surface attributes. Our method can also be used for oblique projector rendering on planar surfaces. In this case, we need to compute the view frustum that warps the rectangular viewport to a smaller quadrilateral inside the viewport. The results from our method do not have the incorrect depth interpolation problem.

Acknowledgments. We wish to thank Jack Snoeyink for referring us to the previous work on minimal enclosing polygons, as well as to Anselmo Lastra and Greg Welch for their many useful suggestions.

Support for this research comes from NSF ITR grant, "Electronic Books for the Tele-immersion Age" and NSF Cooperative Agreement No. ASC-8920219: "NSF Science and Technology Center for Computer Graphics and Scientific Visualization."

References

[Aggarwal et al. 98] Alok Aggarwal, J. S. Chang, and Chee K. Yap. "Minimum Area Circumscribing Polygons." *The Visual Computer: International Journal of Graphics* 1 (1985), 112–117.

[Berg et al. 00] Mark de Berg, Marc van Kreveld, Mark Overmars, and Otfried Schwarzkopf. *Computationaal Geometry: Algorithms and Applications*, Second edition. Berlin: Springer-Verlag, 2000.

[Faugeras 93] Olivier Faugeras. *Three-Dimensional Computer Vision*. Cambridge, MA: MIT Press, 1993.

[Foley et al. 90] James D. Foley, Andries van Dam, Steven K. Feiner, and John F. Hughes. *Computer Graphics: Principles and Practice*, Second edition. Reading, MA: Addison Wesley, 1990.

[Hoff 98] Kenneth E. Hoff. *Understanding Projective Textures*. Available from World Wide Web (http://www.cs.unc.edu/~hoff/techrep/proj textures.html), 1998.

[O'Rourke 98] Joseph O'Rourke. *Computational Geometry in C*, Second edition. Cambridge, UK: Cambridge University Press, 1998.

[Press et al. 93] William H. Press, Saul A. Teukolsky, William T. Vetterling, and Brian P. Flannery. *Numerical Recipes in C: The Art of Scientific Computing*, Second edition. Cambridge, UK: Cambridge University Press, 1993.

[Raskar et al. 98] Ramesh Raskar, Matthew Cutts, Greg Welch, and Wolfgang Stuerzlinger. "Efficient Image Generation for Multiprojector and Multisurface Displays." In *Proceedings of the 9th Eurographics Workshop on Rendering*, pp. 139–144. Berlin: Springer-Verlag, 1998.

[Raskar 99]   Ramesh Raskar. "Oblique Projector Rendering on Planar Surfaces for a Tracked User." SIGGRAPH Sketch. Available from World Wide Web (http://www.cs.unc.edu/~raskar/Oblique/oblique.pdf), 1999.

[Segal et al. 92]   Mark Segal, Carl Korobkin, Rolf van Widenfelt, Jim Foran, and Paul E. Haeberli. "Fast Shadows and Lighting Effects Using Texture Mapping." *Proc. SIGGRAPH '92, Computer Graphics* 26:2 (1992), 249–252.

[Strang 88]   Gilbert Strang. *Linear Algebra and Its Applications*, Third edition. San Diego, CA: Saunders Publishing/Harcourt Brace Jovanovich College Publishers, 1998.

[Trucco, Verri 98]   Emanuele Trucco and Allessandro Verri. *Introductory Techniques for 3-D Computer Vision.* Englewood, NJ: Prentice Hall, 1998.

[Williams 78]   Lance Williams. "Casting Curved Shadows on Curved Surfaces." *Proc. SIGGRAPH '78, Computer Graphics* 12:3 (1978), 270–274.

[Woo et al. 99]   Mason Woo, Jackie Neider, Tom Davis, and Dave Shreiner (OpenGL Architecture Review Board). *OpenGL Programming Guide, Third Edition: The Official Guide to Learning OpenGL, Version 1.2*, Reading, MA: Addison Wesley, 1999.

Web Information:

http://www.acm.org/jgt/papers/LowIlie03

Kok-Lim Low, Department of Computer Science, CB # 3175, Sitterson Hall, UNC-Chapel Hill, Chapel Hill, NC 27599-3175 (lowk@cs.unc.edu)

Adrian Ilie, Department of Computer Science, CB # 3175, Sitterson Hall, UNC-Chapel Hill, Chapel Hill, NC 27599-3175 (adyilie@cs.unc.edu)

Received May 6, 2002; accepted in revised form April 15, 2003.

Current Contact Information:

Kok-Lim Low, Department of Computer Science, CB # 3175, Sitterson Hall, UNC-Chapel Hill, Chapel Hill, NC 27599-3175 (lowk@cs.unc.edu)

Adrian Ilie, Department of Computer Science, CB # 3175, Sitterson Hall, UNC-Chapel Hill, Chapel Hill, NC 27599-3175 (adyilie@cs.unc.edu)

Vol. 6, No. 4: 29–40

Fast and Accurate Parametric Curve Length Computation

Stephen Vincent
Adobe Systems

David Forsey
Radical Entertainment

Abstract. This paper describes a simple technique for evaluating the length of a parametric curve. The technique approximates sections of the curve to the arc of a circle using only sample points on the curve; no computation of derivatives or other properties is required. This method is almost as quick to implement and compute as chord-length summation, but it is much more accurate. It also provides recursion criteria for adaptive sampling. An extension of the algorithm to estimate the area of a parametric surface is discussed briefly.

1. Introduction

The simplest, fastest, and perhaps most widely-used technique for determining the length of a parametric curve is chord-length summation [Mortenson 85]. Other standard approaches include recursive summation, such as the DeCastlejeu algorithm for Bézier curves [Gravesen 95], or integrating the arc length function either by some quadrature technique or by recasting the problem in the form of integrating a differential equation using a technique such as fourth-order Runge-Kutta.

45

This paper presents a new technique that has the following properties:

- easy to implement;

- cheap to compute;

- has good worst-case accuracy characteristics;

- applies to any parametric curve;

- only requires evaluation of points on the curve.

The technique can be summarized as follows: Given points precomputed along the curve, approximate successive triples of points by circular arcs, the lengths of which can be computed cheaply. For the basic algorithm, sum the length of nonoverlapping arcs; alternatively, sum overlapping arcs to get a second length estimate. Regions of high curvature can readily be detected, and the algorithm can be made recursive as necessary.

Test results show that the accuracy of the method increases by the fourth power of the number of points evaluated; thus, doubling the work will result in a 16-fold reduction in the error. Typically, this means that evaluating 50 points along the curve will result in an error of a few parts in 108, representing an improvement of three orders of magnitude over that obtained by chord-length summation of distances between the same sample points along the curve.

2.   Estimation of Arc Length

Consider three points along a curve: $P0$, $P1$, and $P2$ (see Figure 1). A circle with radius r and center C can be fitted to these points. Let $D1$ be the distance from $P0$ to $P2$, and $D2$ be the sum of the distances from $P0$ to $P1$ and from $P1$ to $P2$.

For a circular arc with angle x, the arc length is rx, and for an isosceles triangle with edges r and peak x, the length of the base is $2r\sin(x/2)$. So we have:

$$\text{Arc length} = r\theta + r\psi;$$
$$D1 = 2r\sin((\theta + \psi)/2);$$
$$D2 = 2r\sin(\theta/2) + 2r\sin(\psi/2).$$

If we assume that $P1$ is approximately equidistant between $P0$ and $P2$, so that θ and ψ are approximately equal, we get:

$$\text{Arc length} = 2r\theta;$$
$$D1 = 2r\sin(\theta);$$
$$D2 = 4r\sin(\theta/2).$$

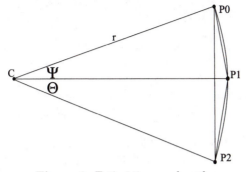

Figure 1. Estimating arc length.

If we also assume that the region has low curvature, then the angles are small, and we can approximate $\sin(x)$ with the first two terms of the Maclaurin series, $\sin(x) = x - x^3/3!$.

Substituting into the above, expanding, and recombining to eliminate r and θ gives:

$$\text{Arc length} = D2 + (D2 - D1)/3.$$

The three-dimensional case reduces to two dimensions simply by working in the plane defined by $P0$, $P1$ and $P2$; the point C will then lie in the same plane.

In the case of a straight line, $D1$ and $D2$ will coincide.

There are two assumptions in the above derivation: first, $P1$ is approximately equidistant from the other two points; second, the angles are small. Geometrically it can be detected if either assumption is invalid; this will be discussed further in Section 4, Awkward Cases.

3.   Basic Algorithm

An odd number of points along the curve are precomputed. Successive length estimation of segments along the curve using the above "circle-arc" approximation provides the total length of the curve.

Precomputation of a suitable number of points (this is discussed later, but 31 would be a reasonable value for a parametric cubic curve) along the curve is essential for the algorithm to function properly: a purely recursive approach cannot be employed. As an example, a curve similar to the one in Figure 3 would be interpreted erroneously as a straight line by a purely recursive approach.

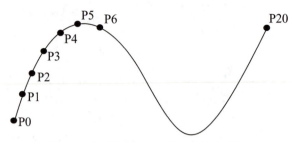

Figure 2. Sliding window estimation.

3.1. Sliding Window Estimation

The sliding window estimation makes two length estimates for each subsection of the curve using the same set of samples. The average length will be a more accurate estimate of the arc length of that segment. As a bonus, it detects regions where our circular approximation assumption is poor by examining the magnitude of the difference between these two estimates.

A high-order error term is associated with the circle-arc ratio calculation; however, any overestimate of the length of a curve segment is cancelled out by a corresponding underestimate of the length of an adjacent segment.

In the example shown in Figure 2, the positions of the 21 samples along the curve are calculated at equal parametric intervals. For all segments, except the first and the last, two estimates of the length are calculated as follows:

1. Consider the segment from $P1$ to $P2$. Using the arc length estimate from $P0$ to $P2$, the arc length from $P1$ to $P2$ will be approximately equal to the ratio of the straight line distance from $P1$ to $P2$ to the sum of the straight line distances from $P0$ to $P1$ and again from $P1$ to $P2$.

2. A second estimate is made by treating the points $(P1, P2, P3)$ as the arc of a circle and applying a similar calculation.

3. If the two estimates are sufficiently close, they are averaged. Estimates that differ by more than a given amount are an indication that a finer sampling of the region is required.

At the start and end segments of the curve, there is only a single estimate for the length of curve in those regions. The length of these segments can be evaluated recursively, or, if the curve is part of a curvature continuous set of curves, by using a subsection that straddles the adjacent curves.

Generally, the sliding window approach improves accuracy by a factor of two over the basic algorithm, which in turn is 200 times more accurate than chord-length summation.

An alternative and simpler approach to the sliding window modification is simply to perform the basic algorithm twice and average the results. The second sampling uses the same set of points as the first, but the indexing of points used for the circle-arc calculation is off by one with respect to the first sample. An appropriate adjustment as above must be made at the ends of the curve for the second sampling. In this approach, we do not check the difference between the estimates of a segment to decide on recursion, but instead we use the same criteria as the basic algorithm. It is this method that is illustrated in the pseudocode.

4. Awkward Cases

There are certain configurations of the shape of a parametric curve that can adversely affect the performance of the basic circle-arc algorithm: points of inflection and regions of high curvature, where the assumption that the arc of a circle can adequately represent a curve segment is no longer valid. Fortunately, detection of these cases is straightforward, and recursive decomposition of the curve in the problem area will yield a more accurate estimate. The notation used follows that used in the description of the algorithm.

Case 1: Points of inflection.   It is possible to have a point of inflection in the middle of a curve segment, which, looks like the curve in Figure 3.

In this case, the distance from $P0$ to $P2$ will equal the sum of the distances from $P0$ to $P1$ and from $P1$ to $P2$. The basic circle-arc algorithm erroneously estimates the curve as a straight line and so underestimates the length of the curve. In practice, the curvature in the region of an inflection point is mild and the resultant error small. If sampled at a reasonable density, little difference in accuracy is apparent between estimates for Bézier curves with and without inflection points. When using the sliding window approach, there will be a significant difference between the two estimates of the length of the curve in the region around the inflection point, indicating that a finer sampling is required to provide an accurate length estimate.

Case 2: Regions of high curvature.   In Figure 4(a), the circle-arc algorithm overestimates the length of the curve. Note, however, that $D2/D1 \gg 1$

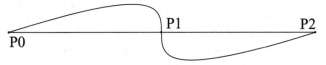

Figure 3. Points of inflection.

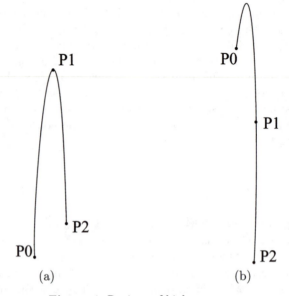

Figure 4. Regions of high curvature.

in such a case, and thus regions of high curvature where a finer step size is required for a better length estimate can be detected.

Figure 4(b) illustrates a variation of this situation. Here the cusp lies close to $P0$ and so $D2$ will approximate D1: i.e., the algorithm will think the curve is a straight line between $P0$ and $P1$ (the x-scale has been exaggerated somewhat for clarity). In this case, the ratio of the two straight-line segments that make up $D2$ will differ significantly from one.

5. Performance Measurements

Performance testing of the circle-arc algorithm used simple curves—such as the ellipse and sin curve—as well as a variety of cubic Bézier curves representative of those found in typical computer graphic applications.

The principle test involved constructing three-dimensional cubic Bézier curves whose control points were random values within the unit cube. With a sample size of 1000, the results obtained for chord-length summation, the basic circle-arc algorithm, and the circle-arc algorithm with the sliding window modification are shown in Figure 5. Results are compared against those obtained using linear chord summation using 100,000 data points. The number of point evaluations refers to the precomputed points only; because of the recursion, the number of points may increase.

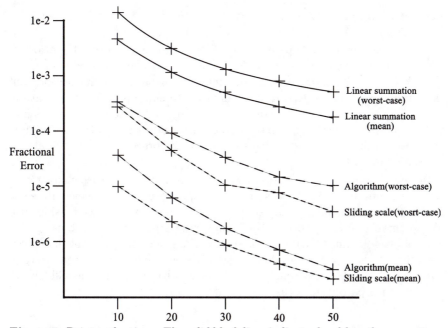

Figure 5. Point evaluations. The solid black lines indicate chord-length summation, the dashed lines the basic arc-length algorithm, and the dotted line the arc-length algorithm with the sliding window modification.

Additional tests used the following curves:

(a) three-dimensional cubic Bézier curves whose first three control points are random values within the unit cube; the fourth control point is (1000,0,0);

(b) two-dimensional cubic Bézier curves with cusps;

(c) two-dimensional cubic Bézier curves with inflection points.

Results for these algorithms were comparable to those obtained for random curves. The results obtained for ellipse and sine curves were equivalent to those for the random cubic Bézier curve tests.

From the data, the error estimate of arc length is inversely proportional to the fourth power of the number of point evaluations. In contrast, chord-length summation has an error inversely proportional to the square of the number of point evaluations.

Vol. 6, No. 4: 29–40

5.1. *Comparison with Quadrature Methods*

We conducted tests to compare the accuracy of this method to integration of the arc-length function using either Simpson's rule or Gauss-Legendre quadrature. For 'typical' curves, using approximately the same amount of work (in the sense of calculating derivatives as opposed to curve points), a quadrature method out-performs the circle-arc approach by at least an order of magnitude. However, Gaussian quadrature in general entails calculation of weights and abscissa at each point, which we assume are precomputed in the above estimation of cost.

However, numerical integration methods have poor worst-case characteristics—behaving badly in the vicinity of a cusp. In such cases, which are easy to detect using a geometric approach, the error in estimating the total length of the curve is up to two orders of magnitude greater than that using the circle-arc algorithm. Guenter and Parent discuss the difficult problem of cusp detection in their paper on adaptive subdivision using Gaussian quadrature [Guenter, Parent 90], but conclude that any quadrature technique will have a problem with pathological cases.

6. Extension to Area Calculations

Initial results indicate that this technique extends well to calculating the area of a parametric surface. An element of the surface can be defined by four points $(u0, u1, v0, v1)$, the area of which $A1$ can be approximated as the sum of two triangles. Recursively subdividing this element into four subelements, and summing the approximate area of these four subelements in the same way, results in a more accurate estimate: $A2$. Combining these two estimates in a way similar to that used for curves yields an estimate for the surface area of $A2 + (A2 - A1)/3$. This seems to give good results, however the curvature characteristics of surfaces are sufficiently complex to require separate coverage.

7. Pseudocode

Sample C code is provided on the web site listed at the end of this paper.

The basic algorithm is presented in two forms—with and without the sliding window modification. In each case, an array of precomputed data points is passed in, but additionally the lower-level GetSegmentLength routine can optionally adaptively recurse and call GetPoint() to evaluate more points on the curve in regions of high curvature.

```
double GetSegmentLength ( t0, t1, t2 ,pt0, pt1, pt2 )
{
    // Compute the length of a small section of a parametric curve
    // From t0 to t2, with t1 being the mid point.
    // The 3 points are precomputed.

    d1 = Distance(pt0,pt2)
    da = Distance(pt0,pt1)
    db = Distance(pt1,pt2)
    d2 = da + db

    // if we're in a region of high curvature, recurse,
    // otherwise return the length as ( d2 + ( d2 - d1 ) / 3 )

    if ( d2<epsilon)
     return ( d2 + ( d2 - d1 ) / 3 )
    else
    {
        if ( d1 < epsilon || d2/d1 > kMaxArc ||
            da < epsilon2 || db/da > kLenRatio ||
            db < epsilon2 || da/db > kLenRatio )
        {
            // Recurse

            /*
            epsilon and epsilon2 in the above expression are there
            to guard against division by zero and underflow. The
            value of epsilon2 should be less than half  that of
            epsilon otherwise unnecessary recursion occurs : values
            of 1e-5 and 1e-6 work satisfactorily.

            kMaxArc implicitly refers to the maximum allowable angle
            that can be subtended by a circular arc : a value of
            1.05 gives good results in practice.

            kLenRatio refers to the maximum allowable ratio in the
            distance between successive data points : a value of 1.2
            gives reasonable results.
            */

            GetPoint ( (t0+t1)/2, mid_pt )
            d3 = GetSegmentLength ( t0, (t0+t1)/2, t1,
                                    pt0, mid_pt, pt1)
            GetPoint ( (t1+t2)/2, mid_pt )
            d4 = GetSegmentLength ( t1, (t1+t2)/2, t2,
                                    pt1, mid_pt, pt2 )
```

```
            return d3 + d4
      }
      else
         return ( d2 + ( d2 - d1 ) / 3 )
   }
}


// The basic algorithm : estimates curve length from min_t to max_t
// sampling the curve at n_pts ( which should be odd ).

CurveLength ( min_t , max_t , n_pts )
{
   P : array of n_pts
   T : corresponding array of t values
   length = 0
   for ( i=0 ; i<n_pts-1; i+=2 )
   {
      length += GetSegmentLength ( T[i],T[i+1],T[i+2],
                                   P[i],P[i+1],P[i+2] )
   }
}


// The above algorithm but with the sliding window modification

CurveLengthSlidingWindow ( min_t , max_t , n_pts )
{
   P : array of n_pts
   T : corresponding array of t values

   // Get first estimate as computed above

   length_a = 0;
   for ( i=0 ; i<n_pts-1; i+=2 )
   {
      length_a += GetSegmentLength ( T[i],T[i+1],T[i+2],
                                     P[i],P[i+1],P[i+2] )
   }
   length_b = 0

   // Start at the second evaluated point : the first will be at min_t

   for ( i=1 ; i<n_pts-2  ; i++ )
   {
      length_b += GetSegmentLength ( T[i],T[i+1],T[i+2],
                                     P[i],P[i+1],P[i+2] )
   }
```

```
// End point corrections : first the first half-section

mid_t = (T[0]+T[1])/2
GetPoint ( mid_t, mid_pt )
length_b += GetSegmentLength ( T[0],(T[0]+T[1])/2,T[1],
                               P[0],mid_pt,P[1] )

// And the last 2 points

mid_t = (T[n_pts-2]+T[n_pts-1])/2
GetPoint ( mid_t, mid_pt )
length_b += GetSegmentLength ( T[n_pts-2],mid_t,T[n_pts-1],
                               P[n_pts-2],mid_pt,P[n_pts-1] )
return ( length_a + length_b ) / 2
}
```

References

[Mortenson 85] Michael E. Mortenson. *Geometric Modeling*. New York: John Wiley & Sons, 1985.

[Gravesen 95] Jens Gravesen. "The Length of Bèzier Curves." In *Graphics Gems V*, edited by Alan Paeth, p.199, San Diego: Academic Press, 1995.

[Guenter, Parent 90] Brian Guenter, Richard Parent. "Computing the Arc Length of Parametric Curves." *IEEE Computer Graphics & Applications*. 10(3): 72–78, 1990.

[Press et al. 92] W.H.Press , S.A. Teukolsky, W.T.Vettering, B.P.Flannery. *Numerical Recipes in C*, 2nd Edition. Cambridge, UK: Cambridge University Press, 1992

Web Information:

Sample C code is provided at http://www.acm.org/jgt/papers/VincentForsey01

Stephen Vincent, Adobe Systems, #155-1800 Stokes St., San Jose, CA, 95126 (svincent@adobe.com)

David Forsey, Radical Entertainment, 8th Floor, 369 Terminal Avenue, Vancouver, BC, Canada, V64 4C4 (dforsey@radical.ca)

Received March 7, 2001; accepted in revised form October 23, 2001.

Vol. 6, No. 4: 29–40

Current Contact Information:

Stephen Vincent, #315-6366 Cassie Ave., Burnaby, B.C., Canada, V5H 2W5 (spvincent@shaw.ca)

David Forsey, Radical Entertainment, 8th Floor, 369 Terminal Avenue, Vancouver, BC, Canada, V64 4C4 (dforsey@radical.ca)

Part II

Polygon Techniques

Vol. 7, No. 1: 13–22

Generalized Barycentric Coordinates on Irregular Polygons

Mark Meyer and Alan Barr
California Institute of Technology

Haeyoung Lee and Mathieu Desbrun
University of Southern California

Abstract. In this paper we present an easy computation of a generalized form of barycentric coordinates for irregular, convex n-sided polygons. Triangular barycentric coordinates have had many classical applications in computer graphics, from texture mapping to ray tracing. Our new equations preserve many of the familiar properties of the triangular barycentric coordinates with an equally simple calculation, contrary to previous formulations. We illustrate the properties and behavior of these new generalized barycentric coordinates through several example applications.

1. Introduction

Mathematicians have known the classical equations for computing triangular barycentric coordinates for centuries. The earliest computer graphics researchers used these equations heavily and produced many useful applications, including function interpolation, surface smoothing, simulation, and ray intersection tests. Due to their linear accuracy, barycentric coordinates can also be found extensively in the finite element literature [Wachpress 75].

Despite the potential benefits, however, it has not been obvious how to generalize barycentric coordinates from triangles to n-sided polygons. Several formulations have been proposed, but most have significant weaknesses. Important properties were lost from the triangular barycentric formu-

Vol. 7, No. 1: 13–22

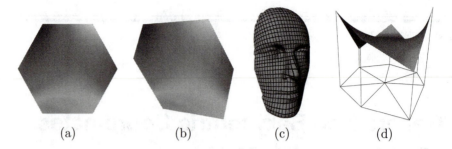

(a) (b) (c) (d)

Figure 1. (a) Smooth color blending using barycentric coordinates for regular polygons [Loop, DeRose 89]. (b) Smooth color blending using our generalization to arbitrary polygons. (c) Smooth parameterization of an arbitrary mesh using our new formula which ensures nonnegative coefficients. (d) Smooth position interpolation over an arbitrary convex polygon (S-patch of depth 1). (See Color Plate XXV.)

lation, which interfered with uses of the previous generalized forms [Pinkall, Polthier 93], [Floater 97]. In other cases, the formulation applied only to regular polygons [Loop, DeRose 89]. Wachspress [Wachpress 75] described an appropriate extension, though it is not very well known in graphics.[1] We will review these techniques and present a much simpler formulation for generalized barycentric coordinates of convex irregular n-gons.

We define the notion of generalized barycentric coordinates in the remainder of this paper as follows: Let $\mathbf{q}_1, \mathbf{q}_2, ..., \mathbf{q}_n$ be n vertices of a *convex* planar polygon Q in $\mathbb{R}^2$, with $n \geq 3$. Further, let $\mathbf{p}$ be an *arbitrary* point inside Q. We define *generalized barycentric coordinates* of $\mathbf{p}$ with respect to $\{\mathbf{q}_j\}_{j=1..n}$ as any set of real coefficients $(\alpha_1, \alpha_2, ..., \alpha_n)$, depending on the vertices of Q and on $\mathbf{p}$, such that all the following properties hold:

- **Property I** (Affine Combination):

$$\mathbf{p} = \sum_{j \in [1..n]} \alpha_j \, \mathbf{q}_j \, , \quad \text{with} \sum_{j \in [1..n]} \alpha_j = 1. \tag{1}$$

  This property allows us to use the polygon's vertices as a basis to locate any point inside. This partition of unity of the coordinates also makes the formulation both rotation and translation invariant.

- **Property II** (Smoothness): The $\{\alpha_j\}_{j=1...n}$ must be infinitely differentiable with respect to $\mathbf{p}$ and the vertices of Q. This ensures smoothness in the variation of the coefficients α_j when we move any vertex $\mathbf{q}_j$.

---

[1]This approach has also been generalized for convex polytopes by Warren [Warren 96].

- **Property III** (Convex Combination):

$$\alpha_j \geq 0 \quad \forall j \in [1 \ldots n].$$

Such a convex combination guarantees no under- or over-shooting in the coordinates: All the coordinates will be between zero and one.

The usual triangular barycentric coordinates ($n = 3$) obviously satisfy the aforementioned properties. Note also that Equation (1) can be rewritten in the following general way:

$$\sum_{j \in [1 \ldots n]} \omega_j (\mathbf{q}_j - \mathbf{p}) = 0. \tag{2}$$

A simple normalization allows us to find partition-of-unity barycentric coordinates:

$$\alpha_j = \omega_j / (\sum_k \omega_k). \tag{3}$$

2. Previous Work

Several researchers have attempted to generalize barycentric coordinates to arbitrary n-gons. Due to the relevance of this extension in CAD, many authors have proposed or used a generalization for *regular n-sided polygons* [Loop, DeRose 89], [Kuriyama 93], [Lodha 93]. Their expressions extend the well-known formula to find barycentric coordinates in a triangle

$$
\begin{aligned}
\alpha_1 &= \mathcal{A}(\mathbf{p}, \mathbf{q}_2, \mathbf{q}_3) / \mathcal{A}(\mathbf{q}_1, \mathbf{q}_2, \mathbf{q}_3) \\
\alpha_2 &= \mathcal{A}(\mathbf{p}, \mathbf{q}_3, \mathbf{q}_1) / \mathcal{A}(\mathbf{q}_1, \mathbf{q}_2, \mathbf{q}_3) \\
\alpha_3 &= \mathcal{A}(\mathbf{p}, \mathbf{q}_1, \mathbf{q}_2) / \mathcal{A}(\mathbf{q}_1, \mathbf{q}_2, \mathbf{q}_3),
\end{aligned}
\tag{4}
$$

where $\mathcal{A}(\mathbf{p}, \mathbf{q}, \mathbf{r})$ denotes the signed area of the triangle $(\mathbf{p}, \mathbf{q}, \mathbf{r})$. Unfortunately, none of the proposed affine combinations leads to the desired properties for irregular polygons. However, Loop and DeRose [Loop, DeRose 89] note in their conclusion that barycentric coordinates defined over arbitrary convex polygons would open many extensions to their work.

Pinkall and Polthier [Pinkall, Polthier 93], and later Eck et al. [Eck et al. 95], presented a conformal parameterization for a triangulated surface by solving a system of linear equations relating the positions of each point $\mathbf{p}$ to the positions of its first ring of neighbors $\{\mathbf{q}_j\}_{j=1 \ldots n}$ as

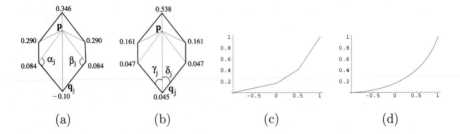

(a)   (b)   (c)   (d)

Figure 2. (a) Polthier's formula [Pinkall, Polthier 93] can give negative coefficients for the barycentric coordinates. (b) Our generalization has guaranteed nonnegativity. (c) A plot of a single barycentric coefficient generated by linearly varying **p** for Floater's method. (d) A plot of a single barycentric coefficient generated by linearly varying **p** for our new method. Note the sharp derivative discontinuities apparent when Floater's method crosses a triangle boundary, while our method produces smoothly varying weights.

$$\sum_j (cot(\alpha_j) + cot(\beta_j))(\mathbf{q}_j - \mathbf{p}) = 0, \tag{5}$$

where the angles are defined in Figure 2(a). As Desbrun et al. demonstrated [Desbrun et al. 99], this formula expresses the gradient of area of the 1-ring with respect to **p**, and therefore Property II is immediately satisfied. The only problem is that the weights can be negative even when the boundary of the n-sided polygon is convex (as indicated in Figure 2(a)), violating Property III.

Floater [Floater 97], [Floater 98] also attempted to solve the problem of creating a parameterization for a surface by solving linear equations. He defined the barycentric coefficients algorithmically to ensure Property III [Floater 97]. Additionally, most of the other properties are also enforced by construction, but due to the algorithmic formulation used, Property II does not hold, as proven by a cross section in Figure 2(c). These barycentric coefficients are only C^0 as the point $\{\mathbf{p}\}$ is moved within the polygon.

However, in 1975, Wachpress proposed a construction of rational basis functions over polygons that leads to the appropriate properties. For the nonnormalized weight ω_j (see Equation (2)) corresponding to the point $\mathbf{q}_j$ of Figure 3(a), Wachspress proposed [Wachpress 75] to use a construction leading to the following formulation:

$$\omega_j = \mathcal{A}(\mathbf{q}_{j-1}, \mathbf{q}_j, \mathbf{q}_{j+1}) \cdot \Pi_{k \notin \{j,j+1\}} \mathcal{A}(\mathbf{q}_{k-1}, \mathbf{q}_k, \mathbf{p}). \tag{6}$$

Thus, each weight ω_j is the product of the area of the jth "boundary" triangle formed by the polygon's three adjacent vertices (shaded in Figure 3), and the areas of the $n - 2$ interior triangles formed by the point **p** and the polygon's adjacent vertices (making sure to exclude the two interior triangles

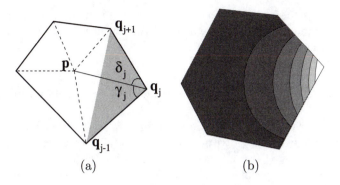

(a) (b)

Figure 3. (a) Our expression for each generalized barycentric coordinates can be computed locally using the edge $\mathbf{pq}_j$ and its two adjacent angles. (b) For an arbitrary convex polygon, the basis function of each vertex is smooth as indicated by the isocontours.

that contain the vertex $\mathbf{q}_j$). The barycentric weights, α_j, are then computed using Equation (3). We refer the reader to Section 2.1 for a very short proof of Property I. Properties II and III obviously stand since we use positive areas for convex polygons, continuous in all the points. In addition to the *pseudoaffine property* defined in [Loop, DeRose 89], this formulation also enforces *edge-preservation*: When $\mathbf{p}$ is on a border edge $\mathbf{q}_i\mathbf{q}_{i+1}$ of the polygon, these barycentric coordinates reduce to the usual linear interpolation between two points.

2.1. Derivation of Equation (6)

We present here a simple proof of Equation (6). As before, let's call $\mathbf{q}_1, ... \mathbf{q}_n$ the n vertices of a convex polygon Q, and $\mathbf{p}$ a point inside Q. If we write the triangular barycentric coordinates for the point $\mathbf{p}$ with respect to a "boundary" triangle $T = (\mathbf{q}_{j-1}, \mathbf{q}_j, \mathbf{q}_{j+1})$, we get (using Equation (4)):

$$\mathcal{A}(T)\,\mathbf{p} = \mathcal{A}(\mathbf{q}_j, \mathbf{q}_{j+1}, \mathbf{p})\,\mathbf{q}_{j-1} + \mathcal{A}(\mathbf{q}_{j-1}, \mathbf{q}_j, \mathbf{p})\,\mathbf{q}_{j+1}$$
$$+ (\mathcal{A}(T) - \mathcal{A}(\mathbf{q}_j, \mathbf{q}_{j+1}, \mathbf{p}) - \mathcal{A}(\mathbf{q}_{j-1}, \mathbf{q}_j, \mathbf{p}))\,\mathbf{q}_j.$$

Since none of these areas can be zero when $\mathbf{p}$ is inside the polygon, we rewrite the previous equation as

$$\frac{\mathcal{A}(T)}{\mathcal{A}(\mathbf{q}_j, \mathbf{q}_{j+1}, \mathbf{p})\mathcal{A}(\mathbf{q}_{j-1}, \mathbf{q}_j, \mathbf{p})}(\mathbf{p} - \mathbf{q}_j)$$
$$= \frac{1}{\mathcal{A}(\mathbf{q}_{j-1}, \mathbf{q}_j, \mathbf{p})}[\mathbf{q}_{j-1} - \mathbf{q}_j] + \frac{1}{\mathcal{A}(\mathbf{q}_j, \mathbf{q}_{j+1}, \mathbf{p})}[\mathbf{q}_{j+1} - \mathbf{q}_j].$$

By summing the contributions of all boundary triangles, the terms on the right hand side of the previous equation will cancel two by two, and we are left with Equation (6).   ∎

3. Simple Computation of Generalized Barycentric Coordinates

In this section, we provide the simplest form of generalized barycentric coordinates for irregular n-sided polygons that retains Property I (affine combination), Property II (smoothness), and Property III (convex combinations) of the triangular barycentric coordinates. This new barycentric formulation is *simple, local, and easy to implement*. In addition, our formulation reduces to the classical equations when $n = 3$, and is equivalent to the analytic form of [Wachpress 75], yet much simpler to compute.

If $\mathbf{p}$ is **strictly** within the polygon Q, we can rewrite the nonnormalized barycentric coordinates given in Equation (6) as

$$
\begin{aligned}
\omega_j &= \frac{\mathcal{A}(\mathbf{q}_{j-1}, \mathbf{q}_j, \mathbf{q}_{j+1})}{\mathcal{A}(\mathbf{q}_{j-1}, \mathbf{q}_j, \mathbf{p})\mathcal{A}(\mathbf{q}_j, \mathbf{q}_{j+1}, \mathbf{p})} \\
&= [sin(\gamma_j + \delta_j)\|\mathbf{q}_j - \mathbf{q}_{j-1}\| \, \|\mathbf{q}_j - \mathbf{q}_{j+1}\|] \, / \\
&\quad [sin(\gamma_j)\|\mathbf{q}_{j-1} - \mathbf{q}_j\| \, \|\mathbf{q}_j - \mathbf{p}\|^2 sin(\delta_j)\|\mathbf{q}_j - \mathbf{q}_{j+1}\|] \\
&= \frac{sin(\gamma_j + \delta_j)}{sin(\gamma_j)sin(\delta_j)\|\mathbf{p} - \mathbf{q}_j\|^2}.
\end{aligned}
$$

Therefore, using trigonometric identities for the sine function, we obtain the following condensed, local formula:

$$
\omega_j = \frac{cot(\gamma_j) + cot(\delta_j)}{\|\mathbf{p} - \mathbf{q}_j\|^2}. \tag{7}
$$

This formulation has the main advantage of being *local*: Only the edge $\mathbf{pq}_j$ and its two adjacent angles, γ_j and δ_j, are needed. The cotangent should not, however, be computed through a trigonometric function call for obvious accuracy reasons. It is far better to use a division between the dot product and the cross product of the triangle involved. Still, compared to the original Equation (6), we obtain locality, hence, simplicity of computation. A simple normalization step to compute the real barycentric coordinates $\{\alpha_j\}_{j=1..n}$ using Equation (3) is the last numerical operation needed. The pseudocode in Figure 4 demonstrates the simplicity of our barycentric formulation when the point $\mathbf{p}$ is strictly within the polygon Q.

Note that the above formulation is valid only when $\mathbf{p}$ is **strictly** within the polygon Q. We remedy this and avoid numerical problems (such as divisions by extremely small numbers) through a special case. If the point $\mathbf{p}$ is within ϵ

// Compute the barycentric weights for a point **p** *in an n-gon Q*
// Assumes **p** *is strictly within Q and the vertices* $\mathbf{q}_j$ *of Q are ordered.*
computeBarycentric(vector2d **p**, polygon Q, int n, real $w[\,]$)
   weightSum $= 0$
   foreach vertex $\mathbf{q}_j$ of Q:
      $prev = (j + n - 1) \bmod n$
      $next = (j + 1) \bmod n$
      $w_j = (\text{cotangent}(\mathbf{p},\mathbf{q}_j,\mathbf{q}_{prev}) + \text{cotangent}(\mathbf{p},\mathbf{q}_j,\mathbf{q}_{next}))/\|\mathbf{p} - \mathbf{q}_j\|^2$
      weightSum $+= w_j$
   // Normalize the weights
   foreach weight w_j:
      $w_j\ /=$ weightSum

// Compute the cotangent of the nondegenerate triangle **abc** *at vertex* **b**
cotangent(vector2d **a**, vector2d **b**, vector2d **c**)
   vector2d **ba** $=$ **a** - **b**
   vector2d **bc** $=$ **c** - **b**
   return $\left(\dfrac{\mathbf{bc}\cdot\mathbf{ba}}{\|\mathbf{bc}\times\mathbf{ba}\|} \right)$

Figure 4. Pseudocode to compute the barycentric weights.

of any of the boundary segments (determined, for instance, by $\|(\mathbf{q}_{j+1} - \mathbf{q}_j) \times (\mathbf{p} - \mathbf{q}_j)\| \leq \epsilon\|\mathbf{q}_{j+1} - \mathbf{q}_j\|)$, the weights can be computed using a simple linear interpolation between the two neighboring boundary points (or even using the nonlocal Equation (6)).

4. Applications

The use of a barycentric coordinate system is extremely useful for a wide range of applications. Since our new formulation easily extends this notion to arbitrary polygons, many domains can benefit from such a simple formula. We describe three very different example applications.

4.1. Interpolation Over n-Sided Polygons

The obvious first application is to use our formula directly for interpolation of any scalar or vector field over polygons. In Figure 1, we demonstrate the smoothness for various six-sided polygons. While the regular case matches with previous formulations [Loop, DeRose 89], [Kuriyama 93], the extension to irregular polygons provides an easy way to guarantee smoothness and nonnegative coefficients.

As mentioned in the introduction, many previous formulations had various shortcomings. In Figure 2, we notice that the Polthier expression [Pinkall, Polthier 93] leads to negative coefficients, while the Floater formulation [Floater 97] only provides $\mathcal{C}^0$ continuity.

4.2. Parameterization

Recently, parameterization of triangular meshes has been studied extensively, focusing on texturing or remeshing of irregular meshes. This consists of defining a piecewise smooth mapping between the triangulated surface and a two-dimensional parameter plane (u, v). Floater [Floater 97], [Floater 98] noticed that a necessary condition to define this mapping is that every vertex of the surface is mapped to a linear combination of its neighbors' mapped positions. It turns out that our formulation provides a new and appropriate way to satisfy this condition. Simply by using Equation (7), we can compute the area-weighted barycentric coordinates of every interior point of a mesh with respect to its 1-ring neighbors directly on the surface. Assuming that no triangles are degenerate, we will obtain a linear combination for every vertex on the surface with respect to its 1-ring neighbors. Solving the resulting linear system of equations (using a conjugate gradient solver to exploit sparsity) will provide a quick and safe way to obtain a parameterization of an arbitrary mesh [Desbrun et al. 02]. Figure 1(c) demonstrates this technique: Starting from an irregular face mesh, we can smoothly map a grid texture on it at low computational cost and in a robust way.

4.3. Surface Modeling and CAD

Loop and DeRose [Loop, DeRose 89] proposed a generalization of Bézier surfaces to regular n-sided polygons, defining what they call S-patches. Relying on a generalization of barycentric coordinates, they unify triangular and tensor product Bézier surfaces. However, they were limited to *regular polygons* with their formulation, which added hard constraints to the modeling process. Similarly, other modeling techniques (see for instance [Volino, Magnenat-Thalmann 98]) use generalized barycentric coordinates, but are again constrained to regular base polygon meshes.

The new formulation we describe provides a very general way to extend the notion of Bézier surfaces. Any convex polygon provides an adequate convex domain to compute a polynomial surface. Figure 1(d) demonstrates the smoothness of a patch of order one, defined over an irregular convex domain, using our barycentric coordinate formula.

5. Conclusion

In this paper, we have introduced a straightforward way to compute smooth, convex-preserving generalized barycentric coordinates. We believe that this simple expression allows many existing works to be extended to irregular polygons with ease. We are currently investigating various avenues, ranging from generalizing this expression to three-dimensional polyhedra, to applying it for the smoothing of meshes.

Acknowledgments. This work has been partially supported by the Integrated Media Systems Center, a NSF Engineering Research Center, cooperative agreement number EEC-9529152, the NSF STC for Computer Graphics and Scientific Visualization (ASC-89-20219), and the CAREER award CCR-0133983.

References

[Desbrun et al. 02] Mathieu Desbrun, Mark Meyer, and Pierre Alliez. "Intrinsic Parameterizations of Surface Meshes." In *Eurographics '02 Proceedings*. Wien: Springer-Verlag, 2002.

[Desbrun et al. 99] Mathieu Desbrun, Mark Meyer, Peter Schröder, and Alan Barr. "Implicit Fairing of Arbitrary Meshes using Laplacian and Curvature Flow." In *Proceedings of ACM SIGGRAPH 99, Computer Graphics Proceedings, Annual Conference Series*, edited by Alyn Rockwood, pp. 317–324, Reading, MA: Addison Wesley Longman, 1999.

[Eck et al. 95] Matthias Eck, Tony DeRose, Tom Duchamp, Hugues Hoppe, Michael Lounsbery, and Werner Stuetzle. "Interactive Multiresolution Surface Viewing." In *Proceedings of SIGGRAPH 95, Computer Graphics Proceedings, Annual Conference Series*, edited by Robert Cook, pp. 91–98, Reading, MA: Addison Wesley, 1995.

[Floater 97] Michael S. Floater. "Parametrization and smooth approximation of surface triangulations." *Computer Aided Geometry Design* 14(3):231–250 (1997).

[Floater 98] Michael S. Floater. "Parametric Tilings and Scattered Data Approximation." *International Journal of Shape Modeling* 4:165–182 (1998).

[Kuriyama 93] S. Kuriyama. "Surface Generation from an Irregular Network of Parametric Curves." *Modeling in Computer Graphics, IFIP Series on Computer Graphics*, pp. 256–274, 1993.

[Loop, DeRose 89] Charles Loop and Tony DeRose. "A multisided generalization of Bézier surfaces." *ACM Transactions on Graphics* 8:204–234 (1989).

[Lodha 93] S. Lodha. "Filling N-sided Holes." In *Modeling in Computer Graphics, IFIP Series on Computer Graphics*, pp. 319–345. Wien: Springer-Verlag, 1993.

[Pinkall, Polthier 93] Ulrich Pinkall and Konrad Polthier. "Computing Discrete Minimal Surfaces and Their Conjugates." *Experimental Mathematics* 2:15–36 (1993).

[Volino, Magnenat-Thalmann 98] Pascal Volino and Nadia Magnenat-Thalmann. "The SPHERIGON: A Simple Polygon Patch for Smoothing Quickly Your Polygonal Meshes." In *Computer Animation '98 Proceedings*, pp. 72–78, 1998.

[Wachpress 75] Eugene Wachpress. *A Rational Finite Element Basis.* manuscript, 1975.

[Warren 96] Joe Warren. "Barycentric Coordinates for Convex Polytopes." *Advances in Computational Mathematics* 6:97–108 (1996).

Web Information:

A simple C++ code implementation is available on the web at http://www.acm.org/jgt/papers/MeyerEtAl02.

Mark Meyer, Department of Computer Science, California Institute of Technology, MS 256-80, Pasadena, CA 91125 (mmeyer@gg.caltech.edu)

Haeyoung Lee, Computer Science Department,University of Southern California, Los Angeles, CA 90089-0781 (leeh@usc.edu)

Alan Barr, Department of Computer Science, California Institute of Technology, MS 256-80, Pasadena, CA 91125 (barr@gg.caltech.edu)

Mathieu Desbrun, Department of Computer Science, University of Southern California, Los Angeles, CA 90089-0781 (desbrun@usc.edu)

Received September 26, 2001; accepted in revised form May 23, 2002.

New Since Original Publication

Since the publication of this paper, the idea of generalized barycentric coordinates has been extensively studied for a variety of applications. In particular, other simple formulas have been proposed in two dimensions by Floater, Hormann, and Kós. Our approach has even been recently extended to three dimensions by Warren, Schaefer, Hirani, and Desbrun, with a similarly simple algorithm. For more details on these new developments, please refer to the additional references.

Additional References

[Floater et al. 05] M. S. Floater, K. Hormann, and G. Kós. "A General Construction of Barycentric Coordinates over Convex Polygons." To appear in *Advances in Computational Mathematics*, 2005.

[Warren et al. 04] Joe Warren, Scott Schaefer, Anil Hirani, and Mathieu Desbrun. "Barycentric Coordinates for nD Smooth Convex Sets." Preprint, 2004.

[Ju et al. 05] Tao Ju, Scott Schaefer, and Joe Warren. "Mean Value Coordinates for Closed Triangular Meshes." To appear in *ACM Trans. on Graphics (SIGGRAPH '05)* (2005).

Current Contact Information:

Mark Meyer, Pixar Animation Studios, 1200 Park Avenue, Emeryville, CA 94608 (mmeyer@pixar.com)

Haeyoung Lee, T719A, 3D-DGP Lab, Department of Computer Science, Hongik University, 72-1 Sangsoo-Dong, Mapo-Gu, Seoul, 121 - 791, Korea (leeh@cs.hongik.ac.kr)

Alan H. Barr, 350 Beckman Institute, MS 350-74, California Institute of Technology, Pasadena, CA 91125 (barr@gg.caltech.edu)

Mathieu Desbrun, MS 158-79, California Institute of Technology, Pasadena, CA 91125 (mathieu@cs.caltech.edu)

Vol. 3, No. 1: 43–46

Computing Vertex Normals from Polygonal Facets

Grit Thürmer and Charles A. Wüthrich

Bauhaus-University Weimar

Abstract. The method most commonly used to estimate the normal vector at a vertex of a polygonal surface averages the normal vectors of the facets incident to the vertex considered. The vertex normal obtained in this way may vary depending on the tessellation of the polygonal surface since the normal of each incident facet contributes equally to the normal in the vertex. To overcome this drawback, we extend the method so that it also incorporates the geometric contribution of each facet, considering the angle under which a facet is incident to the vertex in question.

1. The Problem: Computing the Vertex Normal

Often in computer graphics, one needs to determine normal vectors at vertices of polygonal surfaces, e.g., for Gouraud or Phong shading [Gouraud 71], [Phong 75]. If the polygonal surface approximates a curved surface, whenever possible, vertex normals are provided from the analytical description of the underlying curved surface. Alternatively, a vertex normal N can be computed directly from the polygonal surface. The method commonly suggested is to average the normals N_i of the n facets incident into the vertex [Gouraud 71]:

$$N = \frac{\sum_{i=1}^{n} N_i}{|\sum_{i=1}^{n} N_i|} \qquad (1)$$

If the polygonal surface approximates a curved surface, the normals are only accurate if the facets have regular shape and the orientation does not change much between adjacent polygons. Applying this method, the resulting

71

Vol. 3, No. 1: 43–46   journal of graphics tools

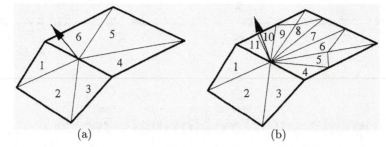

Figure 1. Computed vertex normal for two different meshes of the same geometric situation.

normal vectors depend on the meshing of the surface since the normal of each incident facet contributes equally to the vertex normal. Consequently, if the meshing changes due, e.g., to adaptive tessellation of a deforming surface, the resulting normals will change.

Consider the example shown in Figure 1. Using Equation (1), the computed normal vector varies depending on the tessellation of the polygonal surface. In Figure 1(a), the facets 4, 5, and 6 contribute three of six normals. In Figure 1(b), the facets 4 to 11 contribute eight of eleven normals, despite the fact that the surfaces are the same.

2. The Solution: Average Weighted by Angle

To obtain a result that depends only on the local geometric situation at the vertex, and not on the meshing, we need to consider the spatial contribution of each facet. To do this, we weight the contribution of the normal of each facet by the incident angle of the corresponding facet into the vertex. In symbols,

$$N = \frac{\sum_{i=1}^{n} \alpha_i N_i}{|\sum_{i=1}^{n} \alpha_i N_i|} \tag{2}$$

where α_i is the angle under which the ith facet is incident to the vertex in question and is computed as the angle between the two edges of the ith facet incident in the vertex.

One might consider using other measures, e.g., the area of the polygons, for the weighting factors α_i. However, only the use of the angle as weighting factor results in a normal vector that depends exclusively on the geometric situation around the vertex. Consider the example of Figure 1 again. Since $\alpha_4 + \alpha_5 + \alpha_6$ in (a) is equal to $\alpha_4 + \alpha_5 + ... + \alpha_{11}$ in (b), the normals computed for (a) and (b) by Equation (2) are identical. Thus, applying Equation (2) leads to results that are independent of the tessellation of the polygonal surface. Moreover, the normal at a vertex stays the same regardless of any additional tessellation of the surface.

The proposed extension can be easily incorporated into implementations based on Equation (1), e.g., as suggested by Glassner [Glassner 94]. Of course, this leads to an increase in processing time, but the additional computations are fast and need to be done only once for a polygonal surface.

For the numerical computation, one must be concerned about division by zero in Equation (2). For some applications, e.g., collision detection, one only needs the direction of the normal, so the normalization step can be skipped. Otherwise, if all normals of incident facets lie in the same hemisphere, their average can never be zero. In general, this is the case if the polygonal surface is two-manifold in the vertex, i.e., the facets incident in the vertex form a circular list with each element intersecting its predecessor and successor in exactly one edge, and no other facets in the list.

Another concern is degeneracies in the mesh, i.e., polygons with no area. If the vertex angle of a polygon is 0°, the weight is zero and the polygon will not contribute to the average. Consequently, the result is not affected. If, however, the vertex angle of a degenerate polygon is 180°, e.g., the apex of a zero-height triangle, there is insufficient information at the vertex to produce the geometrically correct result. Such degeneracies should be avoided. Note that the traditional method, Equation (1), would also produce an incorrect result, although a different one.

Other applications that compute vertex properties from polygonal facets could also potentially benefit from weighting by angle. For example, this technique could be used to compute the radiosity value in vertices from incident elements [Cohen, Greenberg 85] in order to reconstruct a continuous radiosity function from a radiosity solution that was performed with constant elements.

References

[Cohen, Greenberg 85] M. F. Cohen and D. P. Greenberg. "The Hemi-cube: A Radiosity Solution for Complex Environments." *Computer Graphics (Proc. SIGGRAPH 85),* 19(3):31–40 (July 1985).

[Glassner 94] A. Glassner. "Building Vertex Normals from an Unstructured Polygon List." In *Graphic Gems IV,* edited by P. Heckbert, pp. 60–73. San Diego: Academic Press, 1994.

[Gouraud 71] H. Gouraud. "Continuous Shading of Curved Surfaces." *IEEE Transactions on Computers,* C-20(6):623–629 (June 1971).

[Phong 75] B. T. Phong. "Illumination for Computer Generated Pictures." *Communications of the ACM,* 18(6):311–317 (June 1975).

Vol. 3, No. 1: 43–46 journal of graphics tools

Web Information:

The web page for this paper is http://www.acm.org/jgt/papers/ThurmerWuthrich98

Grit Thürmer, CoGVis/MMC[1] Group, Faculty of Media, Bauhaus-University Weimar, 99421 Weimar, Germany (thuermer@medien.uni-weimar.de)

Charles A. Wüthrich, CoGVis/MMC Group, Faculty of Media, Bauhaus-University Weimar, 99421 Weimar, Germany (caw@medien.uni-weimar.de)

Received August 31, 1998; accepted in revised form October 15, 1998

New Since Original Publication

After publishing this paper, we were advised that the following work considers the same problem:
C. H. Séquin. "Procedural Spline Interpolation in UNICUBIX." In *Proceedings of the 3rd USENIX Computer Graphics Workshop, Monterey CA*, pp. 63–83. Berkeley, CA: USENIX, 1986."

Current Contact Information:

Grit Thürmer, Institute of Water Management, Hydraulic Engineering and Ecology GmbH, Freiherr-Vom-Stein-Allee 5, 99425 Weimar (grit.thuermer@iwsoe.de)

Charles A. Wüthrich, CoGVis/MMC Group, Faculty of Media, Bauhaus-University Weimar, 99421 Weimar, Germany (caw@medien.uni-weimar.de)

---

[1]Computer Graphics, Visualization/Man-Machine Communication

Vol. 4, No. 2: 1–6

Weights for Computing Vertex Normals from Facet Normals

Nelson Max

Lawrence Livermore National Laboratory

Abstract. I propose a new equation to estimate the normal at a vertex of a polygonal approximation to a smooth surface, as a weighted sum of the normals to the facets surrounding the vertex. The equation accounts for the difference in size of these facets by assigning larger weights for smaller facets. When tested on random cubic polynomial surfaces, the equation is superior to other popular weighting methods.

1. Introduction

When a surface is approximated by polygonal facets, often only the vertex positions are known, but associated vertex normals are required for smooth shading. Sometimes the model is just a polyhedron, and there is no single correct normal at a vertex. However, in other cases, there is an underlying smooth surface approximated by the facets, whose true normals are to be estimated from the facet geometry. Examples are range data, or measured vertices on a plaster character model. A vertex normal is usually taken as a weighted sum of the normals of facets sharing that vertex. Gouraud [Gouraud 71] suggested equal weights, and Thürmer and Wüthrich [Thürmer, Wüthrich 98][1] propose weighting by the facet angles at the vertex. Here, I propose a new set of weights which help to handle the cases when the facets surrounding a vertex differ greatly in size. The weights are appropriate to, and tested for, the situation when the smooth surface is represented locally near the vertex by a height field approximated by a Taylor series polynomial.

---

[1]See also pages 71–74 in this book.

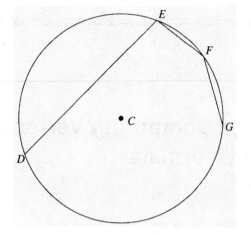

Figure 1. A polygon with unequal sides inscribed in a circle.

The problem of facets of different sizes is illustrated in a two-dimensional analogue in Figure 1, where a circle is approximated by an inscribed polygon $DEFG...$ with unequal sides. The normal to the circle at E passes through the center C of the circle. An unweighted average of the normals to segments DE and EF will lie on the angle bisector of angle DEF, which does not pass through C. The correct weights are inversely proportional to the lengths of the segments. This can be shown by solving simultaneous linear equations, as done in the three-dimensional case below.

2.   The New Weights

The weights I propose in three-dimensions give the correct normals for a polyhedron inscribed in a sphere. Suppose Q is a vertex of the polyhedron, and its adjacent vertices are $V_0, V_1 \ldots, V_{n-1}$. The center C of the sphere can be found as the intersection of the perpendicular bisector planes of the edges QV_i. For simplicity in the derivation, translate the coordinates so that Q is at the origin. Then the midpoint of the edge between $Q = (0,0,0)$ and $V_i = (x_i, y_i, z_i)$ is $(1/2)V_i$, and the vector V_i is normal to any plane perpendicular to this edge, so the equation of the perpendicular bisector plane is

$$x_i x + y_i y + z_i z = \frac{|V_i|^2}{2}.$$

We first consider the case $n = 3$, where we get three linear equations in three unknowns:

$$x_0 x + y_0 y + z_0 z = \frac{|V_0|^2}{2}.$$

$$x_1 x + y_1 y + z_1 z = \frac{|V_1|^2}{2}.$$

$$x_2 x + y_2 y + z_2 z = \frac{|V_2|^2}{2}.$$

Rearranging and grouping the terms of the Cramer's rule solution [MacLane, Birkhoff 67] of this system, one gets

$$C = (x, y, z) = \frac{(|V_2|^2 V_0 \times V_1 + |V_0|^2 V_1 \times V_2 + |V_1|^2 V_2 \times V_0)}{2D}$$

where D is the determinant of the coefficient matrix, and is negative if the vertices V_0, V_1, and V_2 are in counterclockwise order, when viewed from Q. Divide this equation by the negative quantity

$$\frac{|V_0|^2 |V_1|^2 |V_2|^2}{2D}$$

to get a vector pointing in the outward normal direction, that is, a positive multiple c of the normal vector:

$$\frac{V_0 \times V_1}{|V_0|^2 |V_1|^2} + \frac{V_1 \times V_2}{|V_1|^2 |V_2|^2} + \frac{V_2 \times V_0}{|V_2|^2 |V_0|^2} = cN \tag{1}$$

In the degenerate case when Q, V_0, V_1, and V_2 are all in the same plane P, D is zero, but since Q, V_0, V_1, and V_2 then all lie on the circle where the plane P intersects the sphere, one can show that $|V_2|^2 V_0 \times V_1 + |V_0|^2 V_1 \times V_2 + |V_1|^2 V_2 \times V_0$ is also zero, so Equation (1) holds with $c = 0$. If V_0, V_1, and V_2 are in clockwise order when viewed from Q, then c is negative.

To express this equation in terms of the facet normals, let N_i be the normal to the ith facet, between V_i and V_{i+1}, with all indices taken mod n (which, for now, is 3), and let α_i be the angle between V_i and V_{i+1}. Then

$$V_i \times V_{i+1} = N_i |V_i||V_{i+1}| \sin \alpha_i$$

so

$$\sum_{i=0}^{2} \frac{V_i \times V_{i+1}}{|V_i|^2 |V_{i+1}|^2} = \sum_{i=0}^{2} \frac{N_i \sin \alpha_i}{|V_i||V_{i+1}|} = cN. \tag{2}$$

Therefore the proposed weight for normal N_i is $\sin \alpha_i / (|V_i||V_{i+1}|)$. The $\sin \alpha_i$ factor is analogous to the α_i factor in [Thürmer, Wüthrich 98] and the reciprocals of the edge lengths are related to the reciprocal segment lengths discussed in the two-dimensional case above. In practice, the left hand sum in Equation (2) is easier to compute, since it involves neither square roots nor trigonometry.

One can show by induction that also, for any n,

$$\sum_{i=0}^{n-1} \frac{V_i \times V_{i+1}}{|V_i|^2 |V_{i+1}|^2} = cN. \tag{3}$$

The induction starts at $n = 3$ with Equation (2). To derive Equation (3) for $n + 1$, just add Equation (3) for n and Equation (2) for the three vertices V_0, V_{n-1}, and V_n. The last term in the sum in Equation (3), involving $V_{n-1} \times V_0$, will cancel the first term in Equation (2), involving $V_0 \times V_{n-1}$, giving Equation (3) for $n + 1$ (with some new multiple c of N).

In degenerate cases, similar to those discussed in [Thürmer, Wüthrich 98], c will be zero, and no reasonable normal can be chosen. In cases like those in the tests below, when (a) the smooth surface is defined locally by a height function $z = f(x, y)$, (b) the vertices V_i are in monotonically increasing counterclockwise order around Q when viewed from above, and (c) the angles from V_i to V_{i+1} are all acute, every term in Equation (3) will have a positive z component, so the sum will be a positive multiple of the upward surface normal. In other cases, the resulting normal may need to be flipped.

3. Discussion and Results

The derivation above applies only to polyhedra inscribed in spheres, and the normal estimate is incorrect even for ellipsoids. Nevertheless, I believe it is superior to other estimates, because it handles cases where edges adjacent to the vertex Q have very different lengths.

To test this belief, for each n between 3 and 9, I constructed 1,000,000 surfaces of the form

$$z = f(x, y) = Ax^2 + Bxy + Cy^2 + Dx^3 + Ex^2y + Fxy^2 + Gy^3,$$

with A, B, C, D, E, F, and G all uniformly distributed pseudo random numbers in the interval $[-0.1, 0.1]$. These are representative of the third order behavior of smooth surfaces, translated to place the vertex Q, where the normal is to be estimated, at the origin, and rotated to make the true normal at Q point along the positive z axis. I then generated the vertices $V_0, V_1, \ldots, V_{n-1}$ in cylindrical coordinates (r, θ, z) by choosing the θ randomly and uniformly in the interval $[0, 2\pi]$, resorting them in increasing order, choosing for each θ an r randomly and uniformly in the interval $[0, 1]$, converting (r, θ) to Cartesian coordinates (x, y), and then setting $z = f(x, y)$. I rejected cases where

n	(a)	(b)	(c)	(d)	(e)	(f)
3	6.468480	7.301575	10.713165	2.976864	7.355423	7.052390
4	4.324776	5.447873	8.538115	2.275241	5.453043	5.216014
5	3.671877	4.644646	7.030462	1.952892	4.959709	4.588069
6	3.308351	4.234035	5.820398	1.783925	4.890262	4.352113
7	3.088986	4.005317	4.869360	1.674386	4.958369	4.253581
8	2.933336	3.835367	4.064310	1.590934	5.007468	4.169861
9	2.809747	3.719183	3.416770	1.534402	5.115330	4.131644

Table 1. RMS errors in degrees for a cubic surface, with all coefficients in $[-0.1, 0.1]$.

consecutive θ differed by more than π, because they violate condition (c) above.

For each accepted sequence $V_0, V_1, \ldots, V_{n-1}$, I computed several average normals, weighting N_i by: (a) the area of triangle QV_iV_{i+1}, (b) one (i.e., an unweighted average), (c) the angle α_i as proposed in [Thürmer, Wüthrich 98], (d) $\sin\alpha_i/(|V_i||V_{i+1}|)$ as proposed here, (e) $1.0/(|V_i||V_{i+1}|)$, and (f) $1.0/\sqrt{(|V_i||V_{i+1}|)}$. I then measured the error angle between these estimated normals and the correct normal, which is the positive z axis. Table 1 gives the root-mean-square error angle in degrees, over the 1,000,000 trials, for each n and each of the six weighting schemes.

The method proposed here does seem superior, and the method proposed in [Thürmer, Wüthrich 98] seems worse than the unweighted average, at least in the situation tested here, where all vertices are on the smooth surface. The method in [Thürmer, Wüthrich 98] was designed to be consistent when polygons are subdivided, possibly by adding a new vertex along a facet edge which might not be on the smooth surface.

Acknowledgments.   This work was performed under the auspices of the U.S. Department of Energy by Lawrence Livermore National Laboratory under contract W-7405-ENG-48. The paper has been improved significantly using suggestions by John Hughes.

References

[Gouraud 71] Henri Gouraud. "Continuous Shading of Curved Surfaces." *IEEE Transactions on Computers,* C-20(6):623–629 (June 1971).

[MacLane, Birkhoff 67] Saunders Mac Lane and Garrett Birkhoff. *Algebra.* New York: Macmillan, 1967.

[Thürmer, Wüthrich 98] Grit Thürmer and Charles Wüthrich. "Computing Vertex Normals from Polygonal Facets." *jgt* (3)1:43–46 (1998).

Vol. 4, No. 2: 1–6

Web Information:

http://www.acm.org/jgt/papers/Max99

Nelson Max, L-650, Lawrence Livermore National Laboratory, 7000 East Avenue, Livermore, CA 94550 (max2@llnl.gov)

Received June 2, 1999; accepted October 7, 1999

New Since Original Publication

In addition to (a), (b), (c), (d), (e), and (f) in Table 1 in the original paper, I computed the average normals, by: (g) the method in Desbrun et al. [Desbrun et al. 99] (see Table 2). This last method does not average the normals, but instead averages the vectors V_i, using the weights $-(\cot \alpha_i + \cot \beta_i)$, where α_i is the angle $QV_{i+1}V_i$ and β_i is the angle $QV_{i-1}V_i$.

Jin et al. [Jin et al. 05] compare methods by using many other sorts of test cases, including the output of the marching tetrahedra algorithm. As in Table 2, the method in [Desbrun et al. 99] seems to perform poorly in most of the tests in [Jin et al. 05]. The tests in [Jin et al. 05] show that the method proposed here does not perform perfectly even on spheres, in spite of the proof above. For vertices placed directly on the sphere, shown in figure 1a of [Jin et al. 05], the small errors can be attributed to the floating point arithmetic, since floats were used instead of doubles. For the marching tetrahedra contour surface for a sphere, shown in figure 3a of [Jin et al. 05], the larger errors can be explained by the fact that the vertices were placed along tetrahedron edges assuming that the radius function varied linearly along these edges, and thus did not lie exactly on the sphere. I thank the authors of [Jin et al. 05] for clarifying for me why the method proposed here did not perform perfectly on spheres in their tests.

Baerentzen and Aanaes [Baerentzen, Aanaes 05] show that, of the methods discussed, the method in [Thürmer, Wüthrich 98] is the only one that can be used directly for signed distance computations.

n	(a)	(b)	(c)	(d)	(e)	(f)	(g)
3	6.468480	7.301575	10.713165	2.976864	7.355423	7.052390	10.802668
4	4.324776	5.447873	8.538115	2.275241	5.453043	5.216014	14.375840
5	3.671877	4.644646	7.030462	1.952892	4.959709	4.588069	18.409295
6	3.308351	4.234035	5.820398	1.783925	4.890262	4.352113	21.170713
7	3.088986	4.005317	4.869360	1.674386	4.958369	4.253581	22.842968
8	2.933336	3.835367	4.064310	1.590934	5.007468	4.169861	23.653422
9	2.809747	3.719183	3.416770	1.534402	5.115330	4.131644	24.257582

Table 2. RMS errors in degrees for a cubic surface, with all coefficients in $[-0.1, 0.1]$.

Additional References

[Baerentzen, Aanaes 05] J. Andreas Baerentzen and Henrik Aanaes. "Signed Distance Computation Using the Angle Weighted Pseudonormal." *IEEE Transactions on Visualization and Computer Graphics* 11(3): 243–253 (2005).

[Desbrun et al. 99] Mathieu Desbrun, Mark Meyer, Peter Schroeder, and Alan Barr. "Implicit Fairing of Irregular Meshes using Diffusion and Curvature Flow." In *Proceedings of SIGGRAPH 99, Computer Graphics Proceedings, Annual Conference Series,* edited by Alyn Rockwood, pp. 317–324, Reading, MA: Addison Wesley Longman, 1999.

[Jin et al. 05] Shuangshuang Jin, Robert Lewis, and David West. "A Comparison of Algorithms for Vertex Normal Computation." *The Visual Computer* 21(1-2): 71–82 (2005).

Current Contact Information:

Nelson Max, Lawrence Livermore National Laboratory, 7000 East Avenue, Livermore, CA 94550 (max2@llnl.gov)

Vol. 7, No. 2: 9–13

Fast Polygon Area and Newell Normal Computation

Daniel Sunday

John Hopkins University Applied Physics Laboratory

Abstract. The textbook formula for the area of an n-vertex two-dimensional polygon uses $2n + 1$ multiplications and $2n - 1$ additions. We give an improved formula that uses $n + 1$ multiplications and $2n - 1$ additions. A similar formula is derived for a three-dimensional planar polygon where, given the unit normal, the textbook equation cost of $6n + 4$ multiplications and $4n + 1$ additions is reduced to $n + 2$ multiplications and $2n - 1$ additions. Our formula also speeds up Newell's method to compute a robust approximate normal for a nearly planar three-dimensional polygon, using $3n$ fewer additions than the textbook formula. Further, when using this method, one can get the polygon's planar area as equal to the length of Newell's normal for a small additional fixed cost.

1. Introduction

We present a fast method for computing polygon area that was first posted on the author's website (www.geometryalgorithms.com), but has not been previously published. It is significantly faster than the textbook formulas for both two-dimensional and three-dimensional polygons. Additionally, this method can be used to speed up computation of Newell's normal for a three-dimensional polygon. Complete source code is available on the website listed at the end of this paper.

2. The Two-Dimensional Case

For a two-dimensional polygon with n vertices, the standard textbook formula for the polygon's signed area [O'Rourke 98] is:

Vol. 7, No. 2: 9–13

$$A = \frac{1}{2}\sum_{i=0}^{n-1}(x_iy_{i+1} - x_{i+1}y_i).$$

Each term in the sum involves two multiplications and one subtraction; summing up all the terms take another $n - 1$ additions, and the division by two brings the total to $2n + 1$ multiplications and $2n - 1$ additions. (This ignores the loop's $n - 1$ increments of i, as do all subsequent counts.)

With a little algebra, this formula can be simplified. We extend the polygon array by defining $x_n = x_0$ and $x_{n+1} = x_1$, and similarly for y. With these definitions, we get:

$$
\begin{aligned}
2A &= \sum_{i=0}^{n-1}(x_iy_{i+1} - x_{i+1}y_i) \\
&= \sum_{i=0}^{n-1}x_iy_{i+1} - \sum_{i=0}^{n-1}x_{i+1}y_i \\
&= \sum_{i=1}^{n}x_iy_{i+1} - \sum_{i=1}^{n}x_iy_{i-1} \\
&= \sum_{i=1}^{n}x_i(y_{i+1} - y_{i-1})
\end{aligned}
$$

where the third equality comes from recognizing that x_0y_1 is the same as x_ny_{n+1} in the first sum, and index shifting in the second sum. Evaluating the final sum takes n multiplications and $2n - 1$ additions, and dividing by two to get the area brings the total to $n + 1$ multiplications and $2n - 1$ additions.

Using arrays with the first two vertices duplicated at the end, example C code would be:

```
double findArea(int n, double x[], double y[]) {
    int i;
    double sum = 0.0;
    for (i=1; i <= n; i++) {
        sum += (x[i] * (y[i+1] - y[i-1]));
    }
    return (sum / 2.0);
}
```

By duplicating vertices at the end, we have removed the need to do modular arithmetic calculations on the indices, another source of efficiency. Although a good optimizing compiler will produce fast runtime code from indexed array computations, explicitly efficient C++ code using pointers is given on the website at the end of this paper.

An alternate implementation technique [O'Rourke 98], useful when the vertices of a polygon can be added or deleted dynamically, is to use a circular, doubly-linked list of vertex points beginning at V_{start}. Then, efficient code could be:

```
double findArea(Vertex* Vstart) {
    double sum = 0.0;
    Vertex *cV, *pV, *nV; // vertex pointers
    for (cV = Vstart; ; cV = nV) {
            pV = cV->prev; // previous vertex
            nV = cV->next; // next vertex
            sum += (cV->x * (nV->y - pV->y));
            if (nV == Vstart) break;
    }
    return (sum / 2.0);
}
```

3. The Three-Dimensional Case

In three dimensions, the standard formula for the area of a planar polygon [Goldman 94] is,

$$A = \frac{\mathbf{n}}{2} \cdot \sum_{i=0}^{n-1} V_i \times V_{i+1},$$

where the $V_i = (x_i, y_i, z_i)$ are the polygon's vertices, $\mathbf{n} = (nx, ny, nz)$ is the unit normal vector of its plane, and the signed area A is positive when the polygon is oriented counterclockwise around $\mathbf{n}$. This computation uses $6n + 4$ multiplications and $4n + 1$ additions. We improve this by using the fact that projecting the polygon to two dimensions reduces its area by a constant factor; for example, projection onto the xy-plane reduces the area by a factor of nz. Then we can use the two-dimensional formula above to compute the projected area and divide by nz to get the unprojected area, i.e., we simply compute:

```
Axy = findArea(n, x, y);
Area3D = Axy / nz;
```

If nz is nearly zero, it is better to project onto one of the other planes. If we project onto the zx- or yz-plane, we have to use the formulas:

```
Area3D = Azx / ny; // where Azx = findArea(n, z, x), or
Area3D = Ayz / nx; // where Ayz = findArea(n, y, z)
```

respectively. Thus, one should find the largest of $|nx|$, $|ny|$, and $|nz|$ and use the computation excluding that coordinate. Since the smallest possible value for this is $1/\sqrt{3}$, the largest factor by which one multiplies is $\sqrt{3}$; hence we lose, at most, that factor in precision of the answer and possible overflow is limited. Given a unit normal, this method uses only $n + 2$ multiplications, $2n - 1$ additions, and a small overhead choosing the component to ignore.

As a side benefit, we can use exactly the same trick to compute the values in Newell's robust formula [Tampieri 92] for the approximate normal of a nearly planar three-dimensional polygon:

```
double nwx = Ayz;
double nwy = Azx;
double nwz = Axy;
```

We call this normal vector, $\mathbf{nw} = (nwx, nwy, nwz)$, with the projected areas as its components, the *Newell normal* of the polygon. In the most efficient possible version, the computations of Ayz, Azx, and Axy could all be done in a single loop, amortizing the cost of the loop indexing. The resulting computation uses $3n + 3$ multiplications and $6n - 3$ additions, which is $3n$ fewer additions than the textbook Newell formula.

For a further side benefit, while computing the Newell normal, one can quickly recover the polygon's planar area at the same time. In fact, this area is exactly equal to the length *nlen* of the Newell normal, since $Area3D = Axy/(nwz/nlen) = (Axy * nlen)/Axy = nlen$. Thus, the polygon area is gotten for the small additional cost of three multiplies, two adds, and one square root needed to compute *nlen*. Finally, just three more divides by *nlen* converts the Newell normal to a unit normal.

Acknowledgments.   I'd like to thank both the reviewer and editor for feedback that greatly simplified this paper's presentation, and also for suggesting the application to Newell's method.

References

[Goldman 94] Ronald Goldman. "Area of Planar Polygons and Volume of Polyhedra." In *Graphics Gems II*, pp. 170–171 edited by J. Arvo, Morgan Kaufmann, 1994.

[O'Rourke 98] Joseph O'Rourke. *Computational Geometry in C, Second Edition.* New York: Cambridge University Press, 1998.

[Tampieri 92] Filippo Tampieri. "Newell's Method for Computing the Plane Equation." In *Graphics Gems III*, pp. 231–232, Academic Press, 1992.

Web Information:

C++ source code for all three algorithms is available at:
http://www.acm.org/jgt/papers/Sunday02

Daniel Sunday, Johns Hopkins University, Applied Physics Laboratory, 11100 Johns
Hopkins Road, Laurel, MD 20723 (dan.sunday@jhuapl.edu)

Received March 19, 2002; accepted August 9, 2002.

New Since Original Publication

For some applications, such as when modifying legacy code with fixed array
sizes, it is not always easy to go back and make two redundant data points in
the array by repeating the first two points. But, since only the first and last
points of the array require computing array indices modulo n and since the
sum must be initialized anyway, a slight modification breaks those two special
case triangles out of the loop. The same number of calculations is required.
This eliminates any data point redundancy while retaining the performance
of the method. The revised code for a two-dimensional polygon is as follows.
(Note: I'd like to thank Robert D. Miller, East Lansing, Michigan for pointing
this out to me.)

```
// return the signed area of a 2D polygon (x[],y[])
inline double
findArea(int n, double *x, double *y)        // 2D n-vertex polygon
{
    // Assume that the 2D polygon's vertex coordinates are
    // stored in (x[0], y[0]), ..., (x[n-1], y[n-1]),
    // with no assumed vertex replication at the end.

    // Initialize sum with the boundary vertex terms
    double sum = x[0] * (y[1] - y[n-1]) + x[n-1] * (y[0] - y[n-2]);

    for (int i=1; i < n-1; i++) {
        sum += x[i] * ( y[i+1] - y[i-1] );
    }
    return (sum / 2.0);
}
```

Current Contact Information:

Daniel Sunday, Johns Hopkins University, Applied Physics Laboratory, 11100 Johns
Hopkins Road, Laurel, MD 20723 (dan.sunday@jhuapl.edu)

Vol. 1, No. 2: 1–4

Triangulating Convex Polygons Having T-Vertices

P. Cignoni and C. Montani
IEI-CNR

R. Scopigno
CNUCE-CNR

Abstract. A technique to triangulate planar convex polygons having T-vertices is described. Simple strip or fan tessellation of a polygon with T-vertices can result in zero-area triangles and compromise the rendering process. Our technique splits such a polygon into one triangle strip and, at most, one triangle fan. The technique is particularly useful in multiresolution or adaptive representation of polygonal surfaces and the simplification of surfaces.

Triangulating a planar convex polygon is a simple but important task. Although modern graphics languages support the planar convex polygon as an elementary geometric primitive, better performances are obtained by breaking down polygons into compact sequences of triangles. The triangle-strip (Figure 1(a)) and triangle-fan methods (Figure 1(b)) are compact and efficient ways to represent the sequences of triangles and are provided as output primitives in most graphics libraries (OpenGL [Neider et al. 93] and, partially, PHIGS [Gaskins 92], for example).

The triangulation becomes a little more complicated when one or more vertices of the polygon are T-vertices, i.e., three or more aligned vertices exist. This problem arises, for example, in the adaptive representation of polygonal surfaces ([Muller, Stark 93], [Haemer, Zyda 91]); one of the simplest algorithms to avoid cracks (small holes between polygons at different levels of resolution) is to move some of the vertices of the lower resolution polygons onto the boundary of the higher ones, thus introducing T-vertices. In these

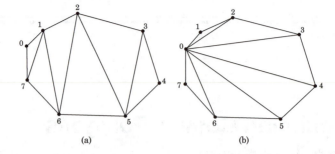

Figure 1. Triangulating a convex polygon: (a) a triangle strip with vertices $v_0, v_7, v_1, v_6, v_2, v_5, v_3, v_4$ and (b) a triangle fan with vertices $v_0, v_7, v_6, v_5, v_4, v_3, v_2, v_1$.

cases, naively creating a triangle strip or fan, as shown in Figure 1, can yield degenerate (zero-area) triangles.

In this article, we present a simple technique which allows us to split convex polygons with T-vertices into one triangle strip and, possibly, one triangle fan. We assume that the user is acquainted with the existence of T-vertices and with their position on the polygon's boundary (this is not a limitation in practical cases), and that the starting vertex of the chain is not a T-vertex (a simple circular shift of the vertices can solve this problem).

The algorithm is shown in Figure 2. Here, the indices *front* and *back* assume the values $0, 1, \cdots$, and $n - 1, n - 2, \cdots$, respectively. Starting from a *regular* vertex (non-T-vertex) of the polygon, the algorithm adds vertices to the triangle strip in the usual zigzag way (Figure 2(a)) until just one unprocessed regular vertex remains. Then, three different cases can occur:

1. The regular vertex is the only vertex to be processed (Figure 2(b)). It can be added as the last point of the strip. There is no need for a triangle fan.

2. The vertex candidate for the insertion into the strip is the regular one (Figure 2(c)). A triangle fan is built. The sequence of vertices of the fan assures the correct triangles' orientation.

3. The vertex candidate for the insertion into the strip is a T-vertex (Figure 2(d) or 2 (e)). The construction of the triangle strip continues until one of the previous situations occurs.

Our simple triangulation is computed in linear time. A Delaunay triangulation of a convex polygon can also be computed in linear time [Aggarwal et al. 89]; our technique is not optimal with respect to the *best* shape of the possible resulting triangles. However, we think the simplicity and effectiveness of our solution makes it valuable in practical cases. The algorithm has been used

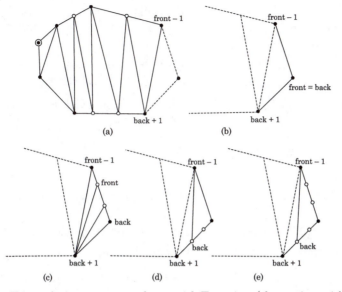

Figure 2. Triangulating a convex polygon with T-vertices (the vertices with a white circle). (a) The construction of the strip interrupts when just one unprocessed regular vertex (the black vertex on the right) is left; (b) if the regular vertex ($v_{front=back}$) is the only vertex to be processed, then it is added as last point of the strip; (c) the next vertex to be processed (v_{back}) is the regular one: a triangle fan is built (in the example the fan $v_{back+1}, v_{back}, \cdots, v_{front}, v_{front-1}$); (d), (e) the candidate vertex (v_{back}) is a T-vertex: it is added to the strip and the algorithm continues until either (b) or (c) occurs.

extensively in the implementation of the DiscMC [Montani et al. 95] [Montani et al. 94], a public domain software for the extraction, simplification, and rendering of isosurfaces from high resolution regular data sets.

References

[Aggarwal et al. 89] A. Aggarwal, L. J. Guibas, J. Saxe, and P. W. Shor. "A Linear Time Algorithm for Computing the Voronoi Diagram of a Convex Polygon." *Discrete and Computational Geometry*, 4:591–604 (1989).

[Gaskins 92] T. Gaskins. *PHIGS Programming Manual*. Sebastopol, CA: O'Reilly & Associates, 1992.

[Haemer, Zyda 91] M. D. Haemer and M. Zyda. "Simplification of Objects Rendered by Polygonal Approximations." *Computers & Graphics*, 15(2):175–184 (1991).

[Montani et al. 94] C. Montani, R. Scateni, and R. Scopigno. "Discretized Marching Cubes." In *Visualization '94 Proceedings* edited by R. Bergeron and A. Kaufman, pp. 281–287. Washington, DC: IEEE Computer Society Press, 1994.

[Montani et al. 95] C. Montani, R. Scateni, and R. Scopigno. "Decreasing Iso-surface Complexity via Discretized Fitting." *Tech. Rep. B4-37-11-95*. Pisa, Italy: IEI-CNR, November 1995.

[Muller, Stark 93] H. Muller and M. Stark. "Adaptive Generation of Surfaces in Volume Data." *The Visual Computer*, 9(4):182–199 (1993).

[Neider et al. 93] J. Neider, T. Davis, and M. Woo. *OpenGL Programming Guide*. Reading, MA: Addison Wesley, 1993.

Web Information:

http://miles.cnuce.cnr.it/cg/homepage.html
DiscMC is available on http://miles.cnuce.cnr.it/cg/swOnTheWeb.html

P. Cignoni, IEI-CNR, Via S. Maria 46, 56126 Pisa, Italy (cignoni@iei.pi.cnr.it)

C. Montani, IEI-CNR, Via S. Maria 46, 56126 Pisa, Italy (montani@iei.pi.cnr.it)

R. Scopigno, CNUCE-CNR, Via S. Maria 36, 56126 Pisa, Italy (r.scopigno@cnuce.cnr.it)

Received March 11, 1996; accepted May 6, 1996

Additional References

Please note that [Montani et al. 95] can now be found at:

[Montani et al. 00] C. Montani, R. Scateni, and R. Scopigno. "Decreasing Iso-surface Complexity via Discrete Fitting." *CAGD* 17(3):207–232 (2000).

Updated Web Information:

The code presented in this paper can now be found at:
http://vcg.isti.cnr.it/jgt/tvert.htm

DiscMC, which is no longer supported, is available at:
http://vcg.isti.cnr.it/downloads/downloads.htm
while more recent open source code for isosurface extraction is available at:
http://vcg.sourceforge.net

Current Contact Information:

P. Cignoni, ISTI-CNR, Via Moruzzi, 1, 56124 Pisa, Italy (p.cignoni@isti.cnr.it)

C. Montani, ISTI-CNR, Via Moruzzi, 1, 56124 Pisa, Italy (c.montani@isti.cnr.it)

R. Scopigno, ISTI-CNR, Via Moruzzi, 1, 56124 Pisa, Italy (r.scopigno@isti.cnr.it)

Vol. 7, No. 4: 69–82

Fast Normal Map Generation for Simplified Meshes

Yigang Wang
Hangzhou Institute of Electronics Engineering, China

Bernd Fröhlich
Bauhaus-Universität Weimar, Germany

Martin Göbel
Fraunhofer Institute for Media Communication, Germany

Abstract. Approximating detailed models with coarse, normal mapped meshes is a very efficient method for real-time rendering of complex objects with fine surface detail. In this paper, we present a new and fast normal map construction algorithm. We scan-convert each triangle of the simplified model, which results in a regularly spaced point set on the surface of each triangle. The original model and all these point samples of the simplified model are rendered from uniformly distributed camera positions. The actual normal map is created by computing the corresponding normals for the point sets. For each point, the normal of the closest point from the high resolution mesh over the set of all camera positions is chosen. Our approach works for general triangle meshes and exploits fully common graphics rendering hardware. Normal map construction times are generally in the range of only a few seconds even for large models. We render our normal-mapped meshes in real-time with a slightly modified version of the standard bump-mapping algorithm. In order to evaluate the approximation error, we investigate the distance and normal errors for normal-mapped meshes. Our investigation of the approximation errors shows that using more than 12 viewpoints does not result in a further improvement in the normal maps for our test cases.

1. Introduction

During the past year, simplification of triangle meshes has been an active area of research in computer graphics. Various groups [Soucy et al. 96], [Cignoni et al. 98], [Sander et al. 00], [Cohen et al. 98], [Sander et al. 01], have shown that texture- and normal-map-based representations can efficiently preserve the geometric and chromatic detail of the original model. Recent graphics cards support these techniques in hardware and allow single pass real-time rendering of complex texture and normal mapped geometry.

Only few methods have been proposed for generating normal maps [Soucy et al. 96], [Cignoni et al. 98], [Sander et al. 00], [Cohen et al. 98], [Sander et al. 01]. These methods usually scan-convert each triangle of the simplified model to obtain a regularly spaced point set. The corresponding normals are computed from the high resolution model and they are packed into a rectangular array—the normal map. The main issue is the construction of a mapping from the simplified model to the original model to calculate the corresponding normal for each point of the normal map. Two approaches have been suggested for solving this problem. One approach directly computes the corresponding normal vector for each point of the normal map [Cignoni et al. 98], [Sander et al. 00]. The other approach assigns the corresponding normal vectors by parameterizations [Cohen et al. 98], [Sander et al. 01]. However, these methods are rather time-consuming for large models.

In this paper, we present a new method, which makes great use of common graphics rendering hardware to accelerate the normal map construction. In our method, the scan-converted point set of the simplified mesh's triangles and the corresponding original model are rendered from uniformly distributed camera positions. The actual normal map is created by computing corresponding normals for the point sets. For each point, the normal of the closest point from the high resolution mesh over the set of all camera positions is chosen. Similar to [Labsik et al. 00], the distance is estimated from Z-buffer entries. A clear limitation of our method is the requirement that all the faces of a mesh need to be seen from at least one viewpoint. There are several solutions to this problems, including resorting back to other more time-consuming methods such as [Cignoni et al. 98] and [Sander et al. 00] for these critical areas. Our method works as a post-process of a simplification. It can be combined with any existing simplification method even if the topology changed during the simplification process.

We have tested our algorithm for various large models and show that high-quality normal maps for complex objects can be generated on standard PC hardware within a few seconds. Our implementation of normal-map-based rendering makes use of Nvidia's GeForce3 and GeForce4 graphics cards, which allow the representation of normal maps with 16 bits per component. Our error analysis indicates that 12 views are in most cases enough for generating high quality normal maps.

2. Creating Normal Maps

Normal maps for height fields can be generated from a single view point (see Figure 1). General meshes require the information from multiple view points from around the model. The normal map construction algorithm can be described in the following way:

```
Normalmap
constructNormalMap( Model simplified_model, Model original_model)
{
    normalmap = simplified_model.scanConvert();
            //scan convert each face and
            //pack points into rectangular array.
    simplified_model.formNormalmapCoordinates();
            //normal map coordinates
            //are computed according to the packing method.
    normalmap.intialize( simplified_model);
            //set the initial values as interpolated values from
            //vertex normal vectors of the simplified model, and
            //set the initial distance value to a threshold.
    original_model.convertNormalToColor();
    for( int viewpoint =0; viewpoint < max_view_num; viewpoint++)
    {
            setTransformation( viewpoint);
            original_model.render(); //with colors mapped from normals
            readZbuffer(zBufferS);
            readColorBuffer(cBufferS);
            for( int i=0; i <normal_map.sampleNumber(); i++)
            {
                vector t = projectToScreen(normalmap(i).position);
                float z = zBufferS(t.x, t.y);
                float dist = fabs (z − t.z);
                if( dist < normalmap(i).dist )
                {
                 normalmap(i).color = cBufferS(t.x, t.y);
                 normalmap(i).dist = dist;
                }
            }
    }
    normalmap.reconvertColorToNormal();
    return normalmap;
}
```

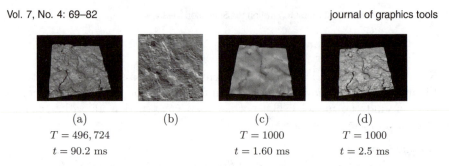

(a)	(b)	(c)	(d)
$T = 496,724$		$T = 1000$	$T = 1000$
$t = 90.2$ ms		$t = 1.60$ ms	$t = 2.5$ ms

Figure 1. Normal maps for height fields mesh can be generated from a single view point. T is the number of triangles of the model; t is the drawing time per frame. (b) Normals are color-coded. (c) Rendering of the simplified model using Gouraud shading on a PC with a Geforce 3 graphics card. (d) Rendering of the simplified model with a normal map on the same PC as (c). (See Color Plate XXIII.)

2.1. Sampling the Simplified Model

We scan-convert each face of the simplified model in software. The triangular sample point sets are packed into a rectangular sample array. We use a simple sampling and packing method. Each face of the simplified mesh is sampled into a right-angled triangle patch of equal resolution, two triangle patches that share one edge in the model form a square, and each of the remaining triangle patches occupies a square. These squares are packed together and form the rectangular normal map array. The actual number of these squares and the user-specified size of the normal map determines the number of samples per square and therefore per simplified triangle. Other more complex sampling and packing approaches can be found in [Cignoni et al. 98] and [Sander et al. 01]. The interpolation within the normal map may result in discontinuous edges, since adjacent sample patches in the packed normal map may be non-adjacent on the simplified mesh. This problem can be avoided with a similar approach chosen in [Cignoni et al. 98] for an antialiasing method. We simply extend each face and save a slightly wider sampling patch (one or two pixels wider in the discrete normal map space) to solve the problem.

2.2. Sampling Viewpoints

For each sampling point on the simplified model, its corresponding point and normal vector on the original model need to be computed. Similar to [Cignoni et al. 98], we define the corresponding point on the original mesh as the one whose distance from the sample point is the shortest. The computation of exact point correspondences based on a distance metric is complicated and time-consuming [Cignoni et al. 98], so we resort to a viewpoint sampling method. For simplicity, we uniformly sample viewpoints around the object.

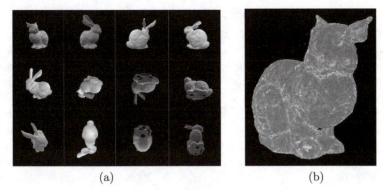

(a)   (b)

Figure 2. (a) The 12 different views of an object used during normal map creation for the bunny model. The object is rendered with colors representing the vertex normal vectors. (b) The normal errors for the bunny model corresponding to a given view. (See Color Plate XXI.)

A set of optimal viewpoint positions can be determined by using Stuerzlinger's [Stuerzlinger 99] hierarchical visibility algorithm.

For each specified camera position, we implement the following four steps:

1. Render the original model onto the screen with vertex colors corresponding to its vertex (or face) normal vectors.

2. Transform each sampling point from the simplified model into current screen coordinates with the same viewing parameters.

3. For each sampling point, replace the corresponding point on the original model (therefore color values) if the distance is smaller than the currently stored distance. For measuring the distance, we just use the difference between the z-values of the sampling point and a point on the original model which maps to the same pixel on the screen.

4. Finally, we reconvert the color vectors to normal vectors, which gives us the normal map.

For the above procedure, we use a fixed set of viewpoints, and require that they are uniformly distributed around the model. For our models, 12 viewpoints were adequate in most cases (see Figure 2(a))—even for objects with complex geometry. In general, the final normal vectors of the normal maps have smooth transitions between two different viewpoints, since the images from neighboring viewpoints typically overlap.

2.3. Improving Normal Vector Precision

The above method usually creates normal maps with 8-bit precision for each component because color buffers typically have three 8-bit color components. There are two kinds of errors, one comes from quantizing the vertex normal to an 8-bit color component. The other is a result of the 8-bit color interpolation. For avoiding these quantization errors as far as possible, we could use an item buffer. Instead of directly rendering the normals of the high resolution model, we render triangle IDs as colors in a first pass. The corresponding point and normal on the original mesh can be found by the triangle ID and the pixel coordinates and depth of the triangle for a given sample point. The quantization of the normal happens then just before it is entered into the normal map, which could be 8- or 16-bit in our case.

2.4. Filling Invisible Parts of the Normal Map

A fixed set of viewpoints is adequate for a wide variety of models. However, for models with very complex visibility, there might be some invisible samples. Often such parts are not important since they are difficult to see if the object is just observed from the outside. However, these undefined samples may result in apparent visible artifacts. In order to avoid undefined areas in the normal map, we initialize the normal map by interpolating the vertex normals of the simplified model. We also provide a distance threshold, and only update the elements in the normal map if the corresponding distance is less than the given threshold. As a consequence, the samples corresponding to invisible parts will keep their initial values, which are the interpolated normal vectors of the simplified model. Thus, we avoid visible artifacts in these regions and use at least Phong's normal interpolation method.

Another basic alternative for avoiding invisible parts in the normal map is to interactively update the normal map by rotating the model in track ball mode. In this way, new views assigned by the user are added to the fixed set of views to improve the normal map.

3. Real-Time Rendering of Normal Mapped Meshes

Single pass real-time rendering of normal-mapped meshes is possible with today's graphics hardware. We use an Nvidia GeForce3 card, which supports vertex programs, texture shaders, and register combiners. Our implementation is pretty similar to a standard real-time bump mapping implementation using Nvidia's extensions, except that we compute the lighting in the local tangent space instead of the texture space.

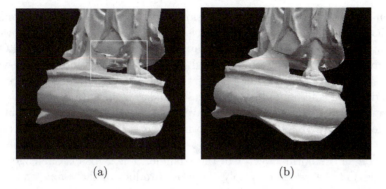

(a)   (b)

Figure 3. Comparison of the methods without and with normal initialization. Both images are synthesized by rendering a simplified model (Buddha) with normal map textures. The left image corresponds to the method without normal initialization, while the right side corresponds to the method with normal initialization. The white rectangle contains some invisible parts for the fixed set of viewpoints, which are below the robe of the statue. (See Color Plate XXII.)

Our experiments showed that an internal precision of 8 bits introduced visual artifacts in high light areas, such as blocky appearance and mach banding. We had to resort to 16-bit precision, which is supported by a texture map type, called GL_SIGNED_HILO_NV. Each element in a GL_SIGNED_HILO_NV texture map consists of two 16-bit signed components HI and LO. HI and LO are mapped to a [-1, 1] range. Such a 2D vector corresponds to a 3D vector (HI, LO, sqrt(1-HI^2 –LO^2)). For this type of texture map, the internal calculations such as dot products are performed with 16-bit precision. Figure 4 compares the results for using these different texture types. We found that the external precision of a normal map is not very important.

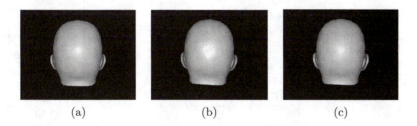

(a)   (b)   (c)

Figure 4. Results for different internal and external normal map precisions. (a) The external normal map type is GL_SIGNED_BYTE; the internal format is GL_SIGNED_HILO_NV. (b) The external normal map type is GL_RGB8; the internal format is the same. (c) The external normal map type is GL_SHORT; the internal format is GL_SIGNED_HILO_NV. (See Color Plate XXIV.)

99

Normal maps with 8- or 16-bit precision lead to very similar visual results if the internal type is set to 16-bit precision computation. In order to use 16-bit internal computations, normal maps have to use GL_SIGNED_HILO_NV texture maps. The normal maps need to be transformed into a space where the z-value of each normal is not less than zero. Our normal maps are filled with discontinuous square patches, which makes it, in most cases, impossible to construct a single consistent texture space for all triangles represented in the normal map. Instead, we transform all normal vectors within a triangle patch of the normal map into a local tangent space, whose z-axis is the same direction as the normal vector of the triangle. The x-axis is one edge of the triangle and the y-axis is simply the cross product of the x-axis and z-axis. As a result, the z-value of the normal vectors in the triangle patch is usually greater than zero, so we can use GL_SIGNED_HILO_NV normal maps and 16-bit computation for our lighting calculations.

4.   Example and Timing Results

We adopted Garland's simplification method [Garland, Heckbert 96] to decimate our models. Our normal map construction and rendering algorithms are implemented under Linux on a 1.3GHz AMD PC with 768M of RAM and 64MB Geforce3 Graphics.

Table 1 shows our construction method as outlined in Section 2. It is interesting to notice that the complexity of the original model does not influence

Model	Number of faces (simplified model)	Construction time (seconds)	Number of samples
Bunny (original model has 69,451 faces)	500	7.39	900,000
	1001	7.26	882,882
	1860	7.75	952,320
	2306	7.51	903,952
	3109	7.51	895,392
	7200	7.7	921,600
Buddha (original model has 1,085,634 faces)	2446	12.23	890,344
	3161	12.07	910,368
	4072	12.18	895,840
	6000	12.03	864,000

Table 1. Normal map construction timings for two models. The normal map size is set to 1024 × 1024. Then normal map construction time depends on the number of points sampled on the simplified model and not very much on the size of the original model.

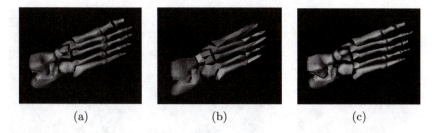

| (a) | (b) | (c) |

Figure 5. Normal maps for a model with simplified topology. (a) Original model. (b) Model with simplified topology. (c) Simplified model with normal map.

the processing time heavily. We found that the rendering time heavily depends on the number of sample points, which need to be transformed by the CPU or through a feedback mechanism of the graphics hardware, which is typically not highly optimized.

An example of a normal-map rendering for a model with simplified topology is shown in Figure 5. Gaps and creases between finger bones disappear in the simplified model. However, these gaps can be seen again in the normal mapped model. In this case, the normals of the original model are well preserved on the simplified model with different topology. This example demonstrates that our method can work well even for complex cases.

Figure 6 shows the rendering of a normal-mapped model on a Geforce3 system. There is an overhead involved compared to conventional Gouraud shaded rendering, but this overhead is usually quite small. For larger simplified models, e.g., with 10,000 or 20,000 triangles, we found that one can become quite quickly geometry limited. This is due to the fact that the length of the vertex program has basically a linear influence on the processing time per vertex. Our vertex program is 42 instructions long. For high resolution renderings—such as 1600×1200—we are usually fill limited. The per-pixel shading computations have here a strong influence.

5. Error Analysis

In order to evaluate the similarity between the original meshes and the normal-mapped simplified meshes, we render images from a set of viewpoints around the model. For our evaluation, we used 30 viewpoints distributed uniformly around the model. For each view, we acquire two images with depth values. The first image is obtained by rendering the original model with vertex colors corresponding to the vertex normals. The second image is obtained by rendering the simplified model with the normal map as a color texture. The depth values allow us to compute distance errors. Normal errors can be obtained

journal of graphics tools

$T = 69,451; t = 14.71$ ms $T = 1000; t = 1.77$ ms $T = 1000; t = 3.54$ ms

$T = 1,085,634; t = 233.14$ ms $T = 3633; t = 2.1$ ms $T = 3633; t = 4.05$ ms

$T = 871,306; t = 169.24$ ms $T = 2997; t = 1.9$ ms $T = 2997; t = 3.8$ ms

Original model   Simplified model   Normal-mapped
                                                  simplified model

Figure 6. Comparison of rendering the original model, a simplified model without normal map, and a simplified model with normal map. Here, T is the triangle count of the model; t is the rendering time per frame.

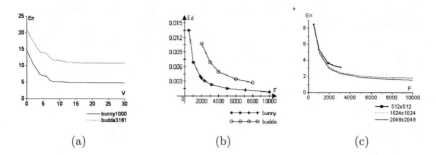

(a)   (b)   (c)

Figure 7. (a) Normal errors versus the number of views (quantized to 8-bit, average vector difference). The normal map resolution is 1024×1024. (b) Distance error versus the face number of simplified models (normalized screen coordinates). (c) Normal errors versus the face number of the simplified mesh and the size of the normal map (bunny model).

from corresponding color values. There are quantization errors involved, since the normal vectors are quantized to 8 bits.

To evaluate the errors versus the number of views, we performed the following experiment. We start with the normal map which is initialized with normal vectors obtained by interpolating the vertex normals of the simplified mesh. The normal map is updated by adding new views one by one. After each update, we compute the normal and distance errors for the current normal-mapped mesh. The result of the experiment is shown in Figure 7(a). It can be seen from the figure that the normal errors cannot be reduced much after 12 views for our models. Another observation is that the normal errors cannot drop far below 5 or 10 for the bunny or Happy Buddha model because of geometrical error.

Note that the normal and position errors of normal-mapped meshes reduce slowly with increasing face numbers. Besides the number of faces, the size of the normal map also influences the visual quality of the normal-mapped mesh. With larger normal map sizes, more normal samples are obtained, which results in less normal errors. (see Figure 7(c)). It is very interesting to notice that it does not make much sense to use very large normal maps for simplified models with a small number of faces, since the geometric error dominates. This basic evaluation gives an idea of how the different parameters affect the quality of the model. However, much more evaluation needs to be performed for a larger variety of models to get more generally applicable conclusions.

In the above, we evaluate the error by averaging all the errors for the entire model. It may be more interesting to see the local errors for each part of the model. To visualize these errors, we render two images for a fixed view as before. One is obtained by rendering the original model with vertex colors corresponding to the vertex normals. The other is obtained by rendering the simplified model with the normal map as a color texture. The difference between the two images is the error for this view. Figure 2(b) shows the error with the appropriate colors. It is clear that the biggest differences appear in detailed parts.

6. Discussion

Sander et al. [Sander et al. 00] mention that the parameterization based on geometrically closest points used in Cignoni et al. [Cignoni et al. 98] leads to discontinuities. One may expect that our method will lead to even more discontinuities. However, we found that the discontinuities are not very obvious with our method, because we use a small number of views. It is clear that there are no discontinuities in areas, which are visible from only one view. Adding further view results in smooth normal transitions in regions seen from multiple views. We extended our method to compute corresponding points by minimizing the angles between interpolated normal vectors and the viewing direction. We found that the quality of the normal map improved only slightly, but the construction process became much slower. These tradeoffs need further study.

Normal vector aliasing is an issue that needs further investigation. Currently, we use only very basic methods to sample the models. Mip-mapping is a well-known technique for antialiasing textures. For meshes that can be represented as $z = f(x, y)$, different mip map levels can be easily generated to avoid normal vector aliasing. However, it is more difficult to extend the idea to the general case. The content of a packed normal maps is not continuous along the triangle boundaries, which means that mip map levels need to be generated by hand by appropriately down sampling the normal information. In addition, our method is limited by the screen resolution at the moment. If the resolution of the original model becomes much larger than the screen resolution, our approach will need to be extended to handle chunks of the model such that the screen resolution is larger than the number of triangles in each chunk. Otherwise, our method down samples the original model by point sampling the normals, which could result in undersampling in this case and lead to aliasing artifacts.

Due to hardware limitations and our single pass rendering, our normal map rendering implementation supports the illumination of textured objects by only a single light source. Support for multiple light sources requires a multipass rendering approach on current graphics hardware.

Our current implementation uses a set of uniformly distributed viewpoints around the model. Stuerzlinger's [Stuerzlinger 99] hierarchical visibility algorithm should be useful to determine a set optimal view positions, which would fully automate our approach and result in even higher quality normal maps.

Acknowledgments. We thank the VRGeo consortium for support and discussions in the context of this work. We also thank the anonymous reviewers who provided excellent feedback on the first version of this paper and helped improve the presentation. The research is also supported by the National Science Foundation of China (No. 60021201). The bunny, dragon, and happy Budda models are provided by the Stanford Computer Graphics Laboratory.

References

[Cignoni et al. 98] P. Cignoni, C. Montani, and R. Scopigno. "A General Method for Preserving Attribute Values on Simplified Meshes." In *Visualization '98 Proceedings*, pp. 59–66. Los Alamitos, CA: IEEE Press, 1998.

[Cohen et al. 98] J. Cohen, M. Olano, and D. Manocha. "Appearance-Preserving Simplification." In *Proceedings of SIGGRAPH 98, Computer Graphics Proceedings, Annual Conference Series*, edited by Michael Cohen, pp. 115–122, Reading, MA: Addison Wesley, 1998.

[Garland, Heckbert 96] M. Garland and P. S. Heckbert. "Surface Simplification Algorithm using Quadric Error Metrics." In *Proceedings of SIGGRAPH 96, Computer Graphics Proceedings, Annual Conference Series*, edited by Holly Rushmeier, pp. 209–216, Reading, MA: Addison Wesley, 1996.

[Garland, Heckbert 98] M. Garland and P. S. Heckbert. "Simplifying Surface with Color and Texture using Quadric Error Metrics." In *IEEE Visualization '98*, pp. 263–269. Los Alamitos, CA: IEEE Press, 1998.

[Heckbert, Garland 97] P. Heckbert and M. Garland. "Survey of Polygonal Surface Simplification Algorithms." In *Multiresolution Surface Modeling (SIGGRAPH '97 Course Notes #25)*. New York: ACM SIGGRAPH, 1997.

[Hoppe 99] Hugues Hoppe. "New Quadric Metric for Simplifying Meshes with Appearance Attributes." In *IEEE Visulization '99 Proceedings*, pp. 59–66. Los Alamitos, CA: IEEE Press, 1999.

[Labsik et al. 00] U. Labsik, R. Sturm, and G. Greiner. "Depth Buffer Based Registration of Free-Form Surfaces." In *VMV 2000 Proceedings*, pp. 121–128. Saarbrücken, Germany: Aka GmbH, 2000.

[Lindstrom, Turk 00] P. Lindstrom and G. Turk. "Image-Driven Simplification." *Transactions on Graphics* 19:3 (2000), 204–241.

[Sander et al. 00] P. V. Sander, X. Gu, S. J. Gortler, H. Hoppe, and J. Snyder. "Silhouette Clipping." In *Proceedings of SIGGRAPH 2000, Computer Graphics Proceedings, Annual Conference Series*, edited by Kurt Akeley, pp. 327–334, Reading, MA: Addison-Wesley, 2000.

Vol. 7, No. 4: 69–82

[Sander et al. 01] P. V. Sander, J. Snyder, S. J. Gortler, and H. Hoppe. "Texture Mapping Progressive Meshes." In *Proceedings of SIGGRAPH 2001, Computer Graphics Proceedings, Annual Conference Series*, edited by E. Fiume, pp. 409–416, Reading, MA: Addison-Wesley, 2001.

[Soucy et al. 96] Marc Soucy, Guy Godin, and Marc Rioux. "A Texture-Mapping Approach for the Compression of Colored 3D Triangulations." *Visual Computer* 12 (1996) 503–514.

[Stuerzlinger 99] W. Stuerzlinger. "Imagine All Visible Surfaces." In *Proceedings of Graphics Interface '99*, pp. 115–122. San Francisco, CA: Morgan Kaufmann Publishers, 1999.

Web Information:

http://www.acm.org/jgt/papers/WangFrohlichGobel02

Yigang Wang, Computer School, Hangzhou Institute of Electronics Engineering, Wenyi Road No. 65, Hangzhou 310037, Zhejiang Province, P.R. China (yumrs@ hzcnc.com)

Bernd Fröhlich, Virtual Reality Systems, Faculty of Media, Bauhaus-Universität Weimar, Bauhausstrasse 11, 99423 Weimar, Germany (bernd.froehlich@medien.uni -weimar.de)

Martin Göbel, Competence Center Virtual Environments, Fraunhofer Institute for Media Communication, Schloss Birlinghoven, D-53754 Sankt Augustin, Germany (martin.goebel@imk.fraunhofer.de)

Received April 2002; accepted September 2002.

Current Contact Information:

Yigang Wang, Department of Computer Science, Hangzhou Dianzi University, Hangzhou Xia'sha University Park, Hangzhou, 310018, P.R.China (wangyg@cad.zju.edu.cn)

Bernd Fröhlich, Virtual Reality Systems, Faculty of Media, Bauhaus-Universität Weimar, Bauhausstrae 11, 99423 Weimar, Germany (bernd.froehlich@medien.uni -weimar.de)

Martin Göbel, fleXilution GmbH, Gottfried-Hagen-Str. 60, 51105 Koeln, Germany (martin.goebel@fleXilution.com)

Part III

Simulation and Collision Detection

Vol. 1, No. 2: 31–50

Fast and Accurate Computation of Polyhedral Mass Properties

Brian Mirtich
University of California at Berkeley

Abstract. The location of a body's center of mass, and its moments and products of inertia about various axes are important physical quantities needed for any type of dynamic simulation or physical based modeling. We present an algorithm for automatically computing these quantities for a general class of rigid bodies: those composed of uniform density polyhedra. The mass integrals may be converted into volume integrals under these assumptions, and the bulk of the paper is devoted to the computation of these volume integrals. Our algorithm is based on a three-step reduction of the volume integrals to successively simpler integrals. The algorithm is designed to minimize the numerical errors that can result from poorly conditioned alignment of polyhedral faces. It is also designed for efficiency. All required volume integrals of a polyhedron are computed together during a single walk over the boundary of the polyhedron; exploiting common subexpressions reduces floating point operations. We present numerical results detailing the speed and accuracy of the algorithm, and also give a complete low level pseudocode description.

1. Introduction

Dynamic simulation of rigid-body systems requires several parameters describing the mass distribution of rigid bodies: the total mass (a scalar), the location of the center of mass (three parameters), and the moments and products of inertia about the center of mass (six parameters). One can always find a *body frame*, with origin at the body's center of mass and axes aligned with its principle axes of inertia, in which the entire mass distribution can be described with a reduced set of four parameters. Nevertheless, the larger parameterization is still needed as a starting point.

109

This paper shows how to efficiently and accurately compute the needed data. The only restrictions are that the body in question be a disjoint union of uniform density polyhedra, given by a boundary representation. We assume one can enumerate over the faces of the polyhedra, and for each face, one can enumerate over the vertices in counterclockwise order. The algorithm is exact, and linear in the number of vertices, edges, or faces of the polyhedra.

The problem of computing mass properties of solid objects has been studied previously. Lee and Requicha give an excellent survey of the various families of algorithms in existence [Lee, Requicha 82]. Our approach is closest to that of Lien and Kajiya, who give an algorithm for computing integrals over arbitrary nonconvex polyhedra, based on a B-rep [Lien, Kajiya 84]. It is $O(n)$ in the polyhedron complexity, and fairly easy to code. In contrast to Lien and Kajiya's algorithm, our algorithm is optimized for computation of mass parameters: it computes all needed mass quantities in parallel during a single traversal of the polyhedra, so that common subexpressions are exploited; it is very fast. In addition, our algorithm adaptively changes the projection direction, thereby reducing numerical errors over those in Lien and Kajiya's and other algorithms.

2. Rigid-Body Mass Parameters

This section defines the rigid-body mass parameters, and their relation to dynamic simulation; readers familiar with these topics may jump to Section 3.. More detailed treatments of this topic may be found in any dynamics text, such as [Greenwood 88] or [Meriam, Kraige 86].

Key quantities in rigid-body dynamics are a body's linear momentum $\mathbf{L}$ and angular momentum $\mathbf{H}$, given by:

$$\mathbf{L} = m\mathbf{v} \qquad (1)$$

$$\mathbf{H} = \mathbf{J}\boldsymbol{\omega}. \qquad (2)$$

Here, $\mathbf{v}$ and $\boldsymbol{\omega}$ are the linear velocity of the center of mass (which we denote $\mathbf{r}$) and the angular velocity of the body, respectively. The scalar m is the mass of the body, and $\mathbf{J}$ is the 3×3 inertia tensor (also called mass matrix) containing the moments and products of inertia:

$$\mathbf{J} = \begin{bmatrix} I_{xx} & -I_{xy} & -I_{xz} \\ -I_{yx} & I_{yy} & -I_{yz} \\ -I_{zx} & -I_{zy} & I_{zz} \end{bmatrix}. \qquad (3)$$

In order to formulate the equations of motion for the body, the quantities m, $\mathbf{J}$, and $\mathbf{r}$ must be determined.

2.1. Computing Mass Parameters with Volume Integrals

The initial problem may be expressed as follows:

Problem 1. Given: *A rigid body comprising N parts, $B_1, \ldots, B_N$, each a uniform density polyhedron. There are no restrictions on the convexity or genus of the polyhedra, nor on the shape of the bounding faces. For each polyhedron B_i, either its density ρ_i or mass m_i is specified, and the geometries of all of the polyhedra are specified relative to a single reference frame. Compute: The mass m, and the reference frame coordinates of the center of mass $\mathbf{r}$ and inertia tensor $\mathbf{J}$ for the entire rigid body.*

The mass m_i and density ρ_i of polyhedral part B_i are related by $m_i = \rho_i V_i$, where V_i is the volume of the polyhedron. Assuming one can compute

$$V_i = \int_{B_i} dV, \tag{4}$$

the masses and densities of each polyhedron can be found. The total mass is $m = \sum_{i=1}^{N} m_i$. The coordinates of the center of mass $\mathbf{r}$ for the entire body are

$$\mathbf{r} = \frac{1}{m} \sum_{i=1}^{N} \rho_i \left(\int_{B_i} x \, dV, \ \int_{B_i} y \, dV, \ \int_{B_i} z \, dV \right)^T. \tag{5}$$

The moments and products of the inertia are given by

$$I'_{xx} = \sum_{i=1}^{N} \rho_i \int_{B_i} (y^2 + z^2) \, dV \tag{6}$$

$$I'_{yy} = \sum_{i=1}^{N} \rho_i \int_{B_i} (z^2 + x^2) \, dV \tag{7}$$

$$I'_{zz} = \sum_{i=1}^{N} \rho_i \int_{B_i} (x^2 + y^2) \, dV \tag{8}$$

$$I'_{xy} = I'_{yx} = \sum_{i=1}^{N} \rho_i \int_{B_i} xy \, dV \tag{9}$$

$$I'_{yz} = I'_{zy} = \sum_{i=1}^{N} \rho_i \int_{B_i} yz \, dV \tag{10}$$

$$I'_{zx} = I'_{xz} = \sum_{i=1}^{N} \rho_i \int_{B_i} zx \, dV. \tag{11}$$

2.2. *Translating Inertias to the Center of Mass*

The inertia quantities in Equations (6)–(11) are primed because they are not exactly the values appearing in the inertia tensor (3). The values in (6)–(11) are computed relative to the given reference frame, but the values in the inertia tensor must be computed relative to a frame with origin at the center of mass. One way to accomplish this is to first compute the location of the center of mass in the given frame, using (5), and then to apply a translation to the body which brings the center of mass to the origin. After performing this transformation, the values computed in (6)–(11) can be directly inserted into the inertia tensor (3).

A better solution is to use the *transfer-of-axis* relations for transferring a moment or product of inertia about one axis to a corresponding one about a parallel axis. To transfer the values computed in (6)–(11) to a frame at the center of mass, one uses (see [Meriam, Kraige 86]):

$$
\begin{array}{llll}
I_{xx} &=& I'_{xx} - m(r_y^2 + r_z^2) \quad (12) & \qquad I_{xy} = I'_{xy} - mr_xr_y \quad (15) \\
I_{yy} &=& I'_{yy} - m(r_z^2 + r_x^2) \quad (13) & \qquad I_{yz} = I'_{yz} - mr_yr_z \quad (16) \\
I_{zz} &=& I'_{zz} - m(r_x^2 + r_y^2) \quad (14) & \qquad I_{zx} = I'_{zx} - mr_zr_x. \quad (17)
\end{array}
$$

The unprimed quantities are inserted into the inertia tensor. If the transfer-of-axis relations are used, one doesn't have to explicitly transform the vertices of the polyhedron after computing the center of mass, hence all of the integrals can be computed simultaneously.

Rigid body dynamics can be computed more efficiently with a reduced set of mass parameters, based on a *body frame*. Computing the body frame amounts to diagonalizing the inertia tensor, a classical problem of linear algebra. The Jacobi method [Press et al. 92] works quite well for this application since **J** is real, symmetric, and of moderate size.

3. Overview of Volume Integration

Equations (4)–(11) show that all required mass properties can be found from ten integrals over volume, for each of the individual polyhedral components. To simplify notation, we drop the polyhedron index and consider only a single polyhedral body. We write the domain of integration as $\mathcal{V}$ as a reminder that it is a volume. The remainder of this paper describes an efficient, exact algorithm for calculating these ten integrals:

$$
T_1 = \int_{\mathcal{V}} 1 \; dV \qquad (18) \qquad\qquad T_x = \int_{\mathcal{V}} x \; dV \qquad (19)
$$

$$T_y = \int_{\mathcal{V}} y \, dV \qquad (20)$$

$$T_z = \int_{\mathcal{V}} z \, dV \qquad (21)$$

$$T_{x^2} = \int_{\mathcal{V}} x^2 \, dV \qquad (22)$$

$$T_{y^2} = \int_{\mathcal{V}} y^2 \, dV \qquad (23)$$

$$T_{z^2} = \int_{\mathcal{V}} z^2 \, dV \qquad (24)$$

$$T_{xy} = \int_{\mathcal{V}} xy \, dV \qquad (25)$$

$$T_{yz} = \int_{\mathcal{V}} yz \, dV \qquad (26)$$

$$T_{zx} = \int_{\mathcal{V}} zx \, dV. \qquad (27)$$

Note that each is an integral of a monomial in x, y, and z. The basic idea is to use the divergence theorem to reduce each of the volume integrals (18)–(27) to a sum of surface integrals over the individual faces of the polyhedron. Each of these surface integrals are evaluated in terms of integrals over a planar projection of the surface. For polygons in the plane, Green's theorem reduces the planar integration to a sum of line integrals around the edges of the polygon. Finally, these line integrals are evaluated directly from the coordinates of the projected vertices of the original polyhedron. Figure 1 illustrates the approach: the volume integral is ultimately reduced to a collection of line integrals in the plane, and the values from these integrations are propagated back into the value of the desired volume integration.

4.   Reduction to Surface Integrals

The first reduction is from an integral over the three-dimensional polyhedral volume to a sum of integrals over its two-dimensional planar faces. This reduction is easily accomplished with the divergence theorem [Stein 82]:

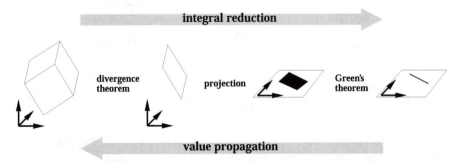

Figure 1. *Strategy for evaluating volume integrals.* Complicated integrals are decomposed into successively simpler ones, and the values from evaluating the simplest integrals are combined and propagated back to evaluate the original ones.

Theorem 1. **(Divergence)** *Let $\mathcal{V}$ be a region in space bounded by the surface $\partial\mathcal{V}$. Let $\hat{\mathbf{n}}$ denote the exterior normal of $\mathcal{V}$ along its boundary $\partial\mathcal{V}$. Then*

$$\int_{\mathcal{V}} \boldsymbol{\nabla} \cdot \mathbf{F} \, dV = \int_{\partial\mathcal{V}} \mathbf{F} \cdot \hat{\mathbf{n}} \, dA \tag{28}$$

for any vector field $\mathbf{F}$ defined on $\mathcal{V}$.

For a polyhedral volume, the right-hand side of (28) can be broken up into a summation over faces of constant normal, so $\hat{\mathbf{n}}$ can be pulled outside the integral. The integrals to be computed, for example $\int_{\mathcal{V}} x \, dV$, do not immediately appear to be of the form in the theorem. But one can find many vector fields whose divergence is the function x; a particularly simple choice is $\mathbf{F}(x,y,z) = (\frac{1}{2}x^2, 0, 0)^T$. This choice has the added advantage that two of its components are identically zero, so that the dot product on the right-hand side of (28) becomes a scalar multiplication. By making similar choices for the other integrals we wish to evaluate, and applying the divergence theorem, Equations (18)–(27) become:

$$T_1 = \sum_{\mathcal{F}\in\partial\mathcal{V}} \hat{n}_x \int_{\mathcal{F}} x \, dA \tag{29} \qquad T_{y^2} = \frac{1}{3}\sum_{\mathcal{F}\in\partial\mathcal{V}} \hat{n}_y \int_{\mathcal{F}} y^3 \, dA \tag{34}$$

$$T_x = \frac{1}{2}\sum_{\mathcal{F}\in\partial\mathcal{V}} \hat{n}_x \int_{\mathcal{F}} x^2 \, dA \tag{30} \qquad T_{z^2} = \frac{1}{3}\sum_{\mathcal{F}\in\partial\mathcal{V}} \hat{n}_z \int_{\mathcal{F}} z^3 \, dA \tag{35}$$

$$T_y = \frac{1}{2}\sum_{\mathcal{F}\in\partial\mathcal{V}} \hat{n}_y \int_{\mathcal{F}} y^2 \, dA \tag{31} \qquad T_{xy} = \frac{1}{2}\sum_{\mathcal{F}\in\partial\mathcal{V}} \hat{n}_x \int_{\mathcal{F}} x^2 y \, dA \tag{36}$$

$$T_z = \frac{1}{2}\sum_{\mathcal{F}\in\partial\mathcal{V}} \hat{n}_z \int_{\mathcal{F}} z^2 \, dA \tag{32} \qquad T_{yz} = \frac{1}{2}\sum_{\mathcal{F}\in\partial\mathcal{V}} \hat{n}_y \int_{\mathcal{F}} y^2 z \, dA \tag{37}$$

$$T_{x^2} = \frac{1}{3}\sum_{\mathcal{F}\in\partial\mathcal{V}} \hat{n}_x \int_{\mathcal{F}} x^3 \, dA \tag{33} \qquad T_{zx} = \frac{1}{2}\sum_{\mathcal{F}\in\partial\mathcal{V}} \hat{n}_z \int_{\mathcal{F}} z^2 x \, dA. \tag{38}$$

5. Reduction to Projection Integrals

Green's theorem reduces an integral over a planar region to an integral around its one-dimensional boundary; however one must start with a region *in the plane*. Although the planar faces of the polyhedron are in three-space, one can project them onto one of the coordinate planes. The next theorem relates integrations over the original face to integrations over the projection.

Theorem 2. *Let $\mathcal{F}$ be a polygonal region in α-β-γ space, with surface normal $\hat{n}$, and lying in the plane*

$$\hat{n}_\alpha \alpha + \hat{n}_\beta \beta + \hat{n}_\gamma \gamma + w = 0. \tag{39}$$

Let Π be the projection of $\mathcal{F}$ into the α-β plane. Then

$$\int_{\mathcal{F}} f(\alpha, \beta, \gamma) \, dA = \frac{1}{|\hat{n}_\gamma|} \int_{\Pi} f(\alpha, \beta, h(\alpha, \beta)) \, d\alpha \, d\beta, \tag{40}$$

where

$$h(\alpha, \beta) = -\frac{1}{\hat{n}_\gamma}(\hat{n}_\alpha \alpha + \hat{n}_\beta \beta + w). \tag{41}$$

Proof. The points $(\alpha, \beta, h(\alpha, \beta))$ lie in the plane of $\mathcal{F}$, so $\mathcal{F}$ is the graph of h over Π. From [Edwards, Penney 86] [Section 17.5, Formula (6)],

$$\int_{\mathcal{F}} f(\alpha, \beta, \gamma) \, dA = \int_{\Pi} f(\alpha, \beta, h(\alpha, \beta)) \sqrt{1 + \left(\frac{\partial h}{\partial \alpha}\right)^2 + \left(\frac{\partial h}{\partial \beta}\right)^2} \, d\alpha \, d\beta. \tag{42}$$

For our h, the square root in the integrand reduces to $|\hat{n}_\gamma|^{-1}$; the theorem follows. $\square$

This theorem provides the reduction of the integral of a polynomial in α, β, and γ over the face $\mathcal{F}$ to the integral of a polynomial in α and β over the projected region Π. From (39), the constant w can be computed: $w = -\hat{n} \cdot \mathbf{p}$, where $\mathbf{p}$ is any point on the face $\mathcal{F}$.

Numerical inaccuracy or floating point errors can occur when the face normal $\hat{n}$ has little or no component in the projection direction; in the extreme situation ($\hat{n}_\gamma = 0$), the face projects to a line segment. To reduce such errors, for a given face the α-β-γ coordinates are always chosen as a right-handed[1] permutation of the of the x-y-z coordinates such that $|\hat{n}_\gamma|$ is maximized. This choice maximizes the area of the projected shadow in the α-β plane (see Figure 2). Note that a choice can always be found such that $|\hat{n}_\gamma| \geq \sqrt{3}^{-1}$.

Recall the integrals we need over the region $\mathcal{F}$ given in (29)–(38). Independent of the three possible correspondences between the x-y-z and α-β-γ coordinates, they all can be found by computing the following 12 integrals:

$$\int_{\mathcal{F}} \alpha \, dA \quad (43) \qquad \int_{\mathcal{F}} \alpha^2 \, dA \quad (46) \qquad \int_{\mathcal{F}} \alpha^3 \, dA \quad (49) \qquad \int_{\mathcal{F}} \alpha^2 \beta \, dA \quad (52)$$

$$\int_{\mathcal{F}} \beta \, dA \quad (44) \qquad \int_{\mathcal{F}} \beta^2 \, dA \quad (47) \qquad \int_{\mathcal{F}} \beta^3 \, dA \quad (50) \qquad \int_{\mathcal{F}} \beta^2 \gamma \, dA \quad (53)$$

$$\int_{\mathcal{F}} \gamma \, dA \quad (45) \qquad \int_{\mathcal{F}} \gamma^2 \, dA \quad (48) \qquad \int_{\mathcal{F}} \gamma^3 \, dA \quad (51) \qquad \int_{\mathcal{F}} \gamma^2 \alpha \, dA. \quad (54)$$

---

[1] We require $\hat{\alpha} \times \hat{\beta} = \hat{\gamma}$.

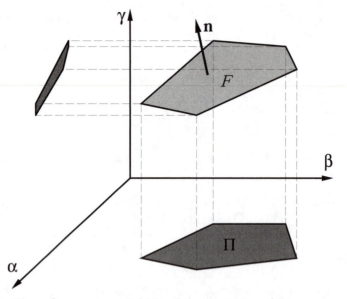

Figure 2. The α-β-γ axes are a right-handed permutation of the x-y-z axes chosen to maximize the size of the face's projected shadow in the α-β plane.

Using Theorem 2, these 12 face integrals can all be reduced to integrals over the projection region Π. For instance,

$$
\begin{aligned}
\int_{\mathcal{F}} \beta^2 \gamma \, dA &= |\hat{n}_\gamma|^{-1} \int_\Pi \beta^2 \frac{\hat{n}_\alpha \alpha + \hat{n}_\beta \beta + w}{-\hat{n}_\gamma} \, d\alpha \, d\beta \\
&= -|\hat{n}_\gamma|^{-1}\hat{n}_\gamma^{-1} \left(\hat{n}_\alpha \int_\Pi \alpha \beta^2 \, d\alpha \, d\beta \; + \; \hat{n}_\beta \int_\Pi \beta^3 \, d\alpha \, d\beta \; + \right. \\
&\qquad \left. w \int_\Pi \beta^2 \, d\alpha \, d\beta \right) \\
&= -|\hat{n}_\gamma|^{-1}\hat{n}_\gamma^{-1} (\hat{n}_\alpha \pi_{\alpha\beta^2} + \hat{n}_\beta \pi_{\beta^3} + w\pi_{\beta^2}),
\end{aligned}
\tag{55}
$$

where

$$
\pi_f = \int_\Pi f \, dA.
\tag{56}
$$

The complete set of face integrals, reduced to projection integrals with Theorem 2, is shown below:

$$
\int_{\mathcal{F}} \alpha \, dA = |\hat{n}_\gamma|^{-1} \pi_\alpha
\tag{57}
$$

$$
\int_{\mathcal{F}} \beta \, dA = |\hat{n}_\gamma|^{-1} \pi_\beta
\tag{58}
$$

$$\int_{\mathcal{F}} \gamma \, dA = -|\hat{n}_\gamma|^{-1} \hat{n}_\gamma^{-1} (\hat{n}_\alpha \pi_\alpha + \hat{n}_\beta \pi_\beta + w \pi_1) \tag{59}$$

$$\int_{\mathcal{F}} \alpha^2 \, dA = |\hat{n}_\gamma|^{-1} \pi_{\alpha^2} \tag{60}$$

$$\int_{\mathcal{F}} \beta^2 \, dA = |\hat{n}_\gamma|^{-1} \pi_{\beta^2} \tag{61}$$

$$\int_{\mathcal{F}} \gamma^2 \, dA = |\hat{n}_\gamma|^{-1} \hat{n}_\gamma^{-2} (\hat{n}_\alpha^2 \pi_{\alpha^2} + 2\hat{n}_\alpha \hat{n}_\beta \pi_{\alpha\beta} + \hat{n}_\beta^2 \pi_{\beta^2} +$$
$$2\hat{n}_\alpha w \pi_\alpha + 2\hat{n}_\beta w \pi_\beta + w^2 \pi_1) \tag{62}$$

$$\int_{\mathcal{F}} \alpha^3 \, dA = |\hat{n}_\gamma|^{-1} \pi_{\alpha^3} \tag{63}$$

$$\int_{\mathcal{F}} \beta^3 \, dA = |\hat{n}_\gamma|^{-1} \pi_{\beta^3} \tag{64}$$

$$\int_{\mathcal{F}} \gamma^3 \, dA = -|\hat{n}_\gamma|^{-1} \hat{n}_\gamma^{-3} (\hat{n}_\alpha^3 \pi_{\alpha^3} + 3\hat{n}_\alpha^2 \hat{n}_\beta \pi_{\alpha^2\beta} + 3\hat{n}_\alpha \hat{n}_\beta^2 \pi_{\alpha\beta^2} +$$
$$\hat{n}_\beta^3 \pi_{\beta^3} + 3\hat{n}_\alpha^2 w \pi_{\alpha^2} + 6\hat{n}_\alpha \hat{n}_\beta w \pi_{\alpha\beta} + 3\hat{n}_\beta^2 w \pi_{\beta^2} +$$
$$3\hat{n}_\alpha w^2 \pi_\alpha + 3\hat{n}_\beta w^2 \pi_\beta + w^3 \pi_1) \tag{65}$$

$$\int_{\mathcal{F}} \alpha^2 \beta \, dA = |\hat{n}_\gamma|^{-1} \pi_{\alpha^2\beta} \tag{66}$$

$$\int_{\mathcal{F}} \beta^2 \gamma \, dA = -|\hat{n}_\gamma|^{-1} \hat{n}_\gamma^{-1} (\hat{n}_\alpha \pi_{\alpha\beta^2} + \hat{n}_\beta \pi_{\beta^3} + w \pi_{\beta^2}) \tag{67}$$

$$\int_{\mathcal{F}} \gamma^2 \alpha \, dA = |\hat{n}_\gamma|^{-1} \hat{n}_\gamma^{-2} (\hat{n}_\alpha^2 \pi_{\alpha^3} + 2\hat{n}_\alpha \hat{n}_\beta \pi_{\alpha^2\beta} + \hat{n}_\beta^2 \pi_{\alpha\beta^2} +$$
$$2\hat{n}_\alpha w \pi_{\alpha^2} + 2\hat{n}_\beta w \pi_{\alpha\beta} + w^2 \pi_\alpha). \tag{68}$$

6. Reduction to Line Integrals

The final step is to reduce an integral over a polygonal projection region in the α-β plane to a sum of line integrals over the edges bounding the region. We adopt the notation as seen in Figure 3. The edges of Π are labeled $\mathcal{E}_1$ through $\mathcal{E}_K$. Edge $\mathcal{E}_e$ is the directed line segment from (α_e, β_e) to $(\alpha_{e+1}, \beta_{e+1})$, $\Delta\alpha_e = \alpha_{e+1} - \alpha_e$, and $\Delta\beta_e = \beta_{e+1} - \beta_e$ [note that $(\alpha_{K+1}, \beta_{K+1}) = (\alpha_1, \beta_1)$]. Finally, edge $\mathcal{E}_e$ has length L_e and exterior unit normal $\hat{\mathbf{m}}_e$.

Green's theorem [2] [Stein 82] provides the final integration reduction:

---

[2]Sometimes more formally called *Green's theorem in the plane*. Additionally, some texts call this *Green's lemma*, reserving *Green's theorem* for a more general 3D result [Wylie, Barrett 82].

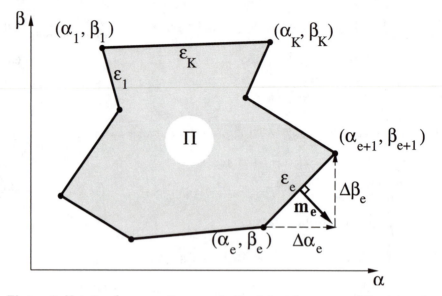

Figure 3. Notation for computing a projection integral as a sum of line integrals.

Theorem 3. (Green's) *Let* Π *be a region in the plane bounded by the single curve* $\partial\Pi$. *Let* $\hat{\mathbf{m}}$ *denote the exterior normal along the boundary. Then*

$$\int_\Pi \boldsymbol{\nabla} \cdot \mathbf{H} \, dA = \oint_{\partial\Pi} \mathbf{H} \cdot \hat{\mathbf{m}} \, ds \qquad (69)$$

for any vector field $\mathbf{H}$ *defined on* Π, *where the line integral traverses the boundary counterclockwise.*

This theorem is a two-dimensional version of the divergence theorem, and our special case again provides simplification. Since Π is polygonal, the right-hand side of (69) may be broken into a summation over edges of constant normal, and by always choosing $\mathbf{H}$ so that one component is identically zero, the dot product becomes a scalar multiplication. From (56) and (57)–(68), all integrals of the form

$$\pi_{\alpha^p\beta^q} = \int_\Pi \alpha^p\beta^q \, dA \qquad (70)$$

are needed for nonnegative integers p and q with $p+q \leq 3$. Consider first the case $q = 0$. By choosing $\mathbf{H} = (\frac{1}{p+1}\alpha^{p+1}, 0)^T$, and applying Green's theorem to the polygonal region Π, we have

$$\int_\Pi \alpha^p \, dA = \frac{1}{p+1}\sum_{e=1}^{K} \hat{m}_{e_\alpha} \int_{\mathcal{E}_e} \alpha(s)^{p+1} \, ds. \qquad (71)$$

In the right-hand integral, the integration variable s is arc length, and runs from 0 to L_e, the length of the edge; $\alpha(s)$ is the α-coordinate of the point on the edge that is a distance s from the starting point. Letting the variable λ be s/L_e, one can change integration variables $(ds = L_e d\lambda)$ to get

$$\int_\Pi \alpha^p \, dA = \frac{1}{p+1} \sum_{e=1}^K \hat{m}_{e_\alpha} L_e \int_0^1 \alpha(L_e\lambda)^{p+1} \, d\lambda. \tag{72}$$

Now $\hat{m}_{e_\alpha} L_e = \pm\Delta\beta_e$, where the plus or minus depends on whether the vertices Π are indexed in counterclockwise or clockwise order, respectively. By convention, we assume that the vertices of the original face $\mathcal{F}$ are indexed in counterclockwise order, thus the vertices of Π will be indexed in counterclockwise order exactly when the sign of $\hat{n}_\gamma$ is positive (Figure 2 helps in visualizing this situation). Hence, $\hat{m}_{e_\alpha} L_e = (\text{sgn}\hat{n}_\gamma)\Delta\beta_e$, and

$$\int_\Pi \alpha^p \, dA = \frac{\text{sgn}\hat{n}_\gamma}{p+1} \sum_{e=1}^K \Delta\beta_e \int_0^1 \alpha(L_e\lambda)^{p+1} \, d\lambda. \tag{73}$$

The case with $q = 1$ is similar, except one chooses $\mathbf{H} = (\frac{1}{p+1}\alpha^{p+1}\beta, 0)^T$. Finally, one can derive analogous equations for the cases when $p = 0$ or $p = 1$. The results are:

$$\int_\Pi \beta^q \, dA = -\frac{\text{sgn}\hat{n}_\gamma}{q+1} \sum_{e=1}^K \Delta\alpha_e \int_0^1 \beta(L_e\lambda)^{q+1} \, d\lambda \tag{74}$$

$$\int_\Pi \alpha^p\beta \, dA = \frac{\text{sgn}\hat{n}_\gamma}{p+1} \sum_{e=1}^K \Delta\beta_e \int_0^1 \alpha(L_e\lambda)^{p+1}\beta(L_e\lambda) \, d\lambda \tag{75}$$

$$\int_\Pi \alpha\beta^q \, dA = -\frac{\text{sgn}\hat{n}_\gamma}{q+1} \sum_{e=1}^K \Delta\alpha_e \int_0^1 \alpha(L_e\lambda)\beta(L_e\lambda)^{q+1} \, d\lambda. \tag{76}$$

7. Evaluation of Integrals from Vertex Coordinates

We have successively reduced the original volume integrals to face integrals, projection integrals, and finally line integrals. The latter can be directly evaluated in terms of vertex coordinates, with the help of the following theorem.

Theorem 4. *Let L_e be the length of the directed line segment from (α_e, β_e) to $(\alpha_{e+1}, \beta_{e+1})$. Let $\alpha(s)$ and $\beta(s)$ be the α- and β-coordinates of the point on this segment a distance s from the starting point. Then for nonnegative integers p and q,*

119

$$\int_0^1 \alpha(L_e\lambda)^p \beta(L_e\lambda)^q \, d\lambda = \frac{1}{p+q+1} \sum_{i=0}^{p} \sum_{j=0}^{q} \frac{\dbinom{p}{i}\dbinom{q}{j}}{\dbinom{p+q}{i+j}} \alpha_{e+1}^i \alpha_e^{p-i} \beta_{e+1}^j \beta_e^{q-j}.$$

(77)

Proof. Denoting the integral on the left-hand side of (77) by I,

$$I = \int_0^1 [(1-\lambda)\alpha_e + \lambda\alpha_{e+1}]^p \, [(1-\lambda)\beta_e + \lambda\beta_{e+1}]^q \, d\lambda \qquad (78)$$

$$= \int_0^1 \left[\sum_{i=0}^{p} B_i^p(\lambda)\alpha_{e+1}^i \alpha_e^{p-i} \right] \left[\sum_{j=0}^{q} B_j^q(\lambda)\beta_{e+1}^j \beta_e^{q-j} \right] d\lambda, \qquad (79)$$

(80)

where the Bs are *Bernstein polynomials*:

$$B_k^n(\lambda) = \binom{n}{k} \lambda^k (1-\lambda)^{n-k}. \qquad (81)$$

Two important properties of these polynomials are [Farouki, Rajan 88], [Farin 90]:

$$B_i^p(\lambda)B_j^q(\lambda) = \frac{\dbinom{p}{i}\dbinom{q}{j}}{\dbinom{p+q}{i+j}} B_{i+j}^{p+q}(\lambda), \qquad (82)$$

$$\int_0^1 B_k^n(\lambda) \, d\lambda = \frac{1}{n+1}. \qquad (83)$$

Expanding the product in (80) and applying (82) gives

$$I = \sum_{i=0}^{p} \sum_{j=0}^{q} \frac{\dbinom{p}{i}\dbinom{q}{j}}{\dbinom{p+q}{i+j}} \alpha_{e+1}^i \alpha_e^{p-i} \beta_{e+1}^j \beta_e^{q-j} \int_0^1 B_{i+j}^{p+q}(\lambda) \, d\lambda. \qquad (84)$$

Evaluating the integrals using (83) proves the theorem. $\square$

All of the line integrals appearing in (73)–(76) are special cases of Theorem 4, with either p or q set to 0 or 1. Specifically,

$$\int_0^1 \alpha^{p+1} \, d\lambda = \frac{1}{p+2} \sum_{i=0}^{p+1} \alpha_{e+1}^i \alpha_e^{p+1-i} \qquad (85)$$

$$\int_0^1 \beta^{q+1}\, d\lambda = \frac{1}{q+2} \sum_{j=0}^{q+1} \beta_{e+1}^j \beta_e^{q+1-j} \tag{86}$$

$$\int_0^1 \alpha^{p+1}\beta\, d\lambda = \frac{1}{(p+2)(p+3)} \left[\beta_{e+1} \sum_{i=0}^{p+1} (i+1)\alpha_{e+1}^i \alpha_e^{p+1-i} + \right.$$
$$\left. \beta_e \sum_{i=0}^{p+1} (p+2-i)\alpha_{e+1}^i \alpha_e^{p+1-i} \right] \tag{87}$$

$$\int_0^1 \alpha\beta^{q+1}\, d\lambda = \frac{1}{(q+2)(q+3)} \left[\alpha_{e+1} \sum_{j=0}^{q+1} (j+1)\beta_{e+1}^j \beta_e^{q+1-j} + \right.$$
$$\left. \alpha_e \sum_{j=0}^{q+1} (q+2-j)\beta_{e+1}^j \beta_e^{q+1-j} \right]. \tag{88}$$

Substituting (85)–(88) into (73)–(76), respectively give all of the needed projection integrals in terms of the coordinates of the projection vertices:

$$\pi_1 = \int_\Pi 1\, dA = +\frac{\mathrm{sgn}\hat{n}_\gamma}{2} \sum_{e=1}^K \Delta\beta_e(\alpha_{e+1} + \alpha_e) \tag{89}$$

$$\pi_\alpha = \int_\Pi \alpha\, dA = +\frac{\mathrm{sgn}\hat{n}_\gamma}{6} \sum_{e=1}^K \Delta\beta_e(\alpha_{e+1}^2 + \alpha_{e+1}\alpha_e + \alpha_e^2) \tag{90}$$

$$\pi_\beta = \int_\Pi \beta\, dA = -\frac{\mathrm{sgn}\hat{n}_\gamma}{6} \sum_{e=1}^K \Delta\alpha_e(\beta_{e+1}^2 + \beta_{e+1}\beta_e + \beta_e^2) \tag{91}$$

$$\pi_{\alpha^2} = \int_\Pi \alpha^2\, dA = +\frac{\mathrm{sgn}\hat{n}_\gamma}{12} \sum_{e=1}^K \Delta\beta_e(\alpha_{e+1}^3 + \alpha_{e+1}^2\alpha_e +$$
$$\alpha_{e+1}\alpha_e^2 + \alpha_e^3) \tag{92}$$

$$\pi_{\alpha\beta} = \int_\Pi \alpha\beta\, dA = +\frac{\mathrm{sgn}\hat{n}_\gamma}{24} \sum_{e=1}^K \Delta\beta_e \left[\beta_{e+1}(3\alpha_{e+1}^2 + 2\alpha_{e+1}\alpha_e + \alpha_e^2) + \right.$$
$$\left. \beta_e(\alpha_{e+1}^2 + 2\alpha_e\alpha_{e+1} + 3\alpha_e^2) \right] \tag{93}$$

121

$$\pi_{\beta^2} = \int_\Pi \beta^2 \, dA \;=\; -\frac{\mathrm{sgn}\hat{n}_\gamma}{12} \sum_{e=1}^{K} \Delta\alpha_e (\beta_{e+1}^3 + \beta_{e+1}^2\beta_e +$$

$$\beta_{e+1}\beta_e^2 + \beta_e^3) \tag{94}$$

$$\pi_{\alpha^3} = \int_\Pi \alpha^3 \, dA \;=\; +\frac{\mathrm{sgn}\hat{n}_\gamma}{20} \sum_{e=1}^{K} \Delta\beta_e (\alpha_{e+1}^4 + \alpha_{e+1}^3\alpha_e + \alpha_{e+1}^2\alpha_e^2 +$$

$$\alpha_{e+1}\alpha_e^3 + \alpha_e^4) \tag{95}$$

$$\pi_{\alpha^2\beta} = \int_\Pi \alpha^2\beta \, dA \;=\; +\frac{\mathrm{sgn}\hat{n}_\gamma}{60} \sum_{e=1}^{K} \Delta\beta_e \left[\beta_{e+1}(4\alpha_{e+1}^3 + 3\alpha_{e+1}^2\alpha_e + \right.$$

$$2\alpha_{e+1}\alpha_e^2 + \alpha_e^3) + \beta_e(\alpha_{e+1}^3 +$$

$$\left. 2\alpha_{e+1}^2\alpha_e + 3\alpha_{e+1}\alpha_e^2 + 4\alpha_e^3) \right] \tag{96}$$

$$\pi_{\alpha\beta^2} = \int_\Pi \alpha\beta^2 \, dA \;=\; -\frac{\mathrm{sgn}\hat{n}_\gamma}{60} \sum_{e=1}^{K} \Delta\alpha_e \left[\alpha_{e+1}(4\beta_{e+1}^3 + 3\beta_{e+1}^2\beta_e + \right.$$

$$2\beta_{e+1}\beta_e^2 + \beta_e^3) + \alpha_e(\beta_{e+1}^3 +$$

$$\left. 2\beta_{e+1}^2\beta_e + 3\beta_{e+1}\beta_e^2 + 4\beta_e^3) \right] \tag{97}$$

$$\pi_{\beta^3} = \int_\Pi \beta^3 \, dA \;=\; -\frac{\mathrm{sgn}\hat{n}_\gamma}{20} \sum_{e=1}^{K} \Delta\alpha_e (\beta_{e+1}^4 + \beta_{e+1}^3\beta_e + \beta_{e+1}^2\beta_e^2 +$$

$$\beta_{e+1}\beta_e^3 + \beta_e^4). \tag{98}$$

8. Algorithm

Based on the derivation in Sections 4–7, we give a complete algorithm for computing the ten desired volume integrals (18)–(27).

The algorithm comprises three routines:

1. CompVolumeIntegrals($\mathcal{V}$) (Figure 4) computes the required volume integrals for a polyhedron by summing surface integrals over its faces, as detailed in Equations (29)–(38).

2. CompFaceIntegrals($\mathcal{F}$) (Figure 5) computes the required surface integrals over a polyhedral face from the integrals over its projection, as detailed in Equations (57)–(68).

3. CompProjectionIntegrals($\mathcal{F}$) (Figure 6) computes the required integrals over a face projection from the coordinates of the projections vertices, as detailed in Equations (89)–(98).

The algorithm contains a slight simplification over the presented derivation. When Equations (89)–(98) are substituted into (57)–(68), the computation of $\mathrm{sgn}\hat{n}_\gamma$ and $|\hat{n}_\gamma|$ becomes unnecessary, since these terms always appear together in a product, giving simply $\hat{n}_\gamma$. Thus, no signs or absolute values are computed in the routines CompFaceIntegrals and CompProjectionIntegrals.

9. Test Results

We now analyze the speed and accuracy of the algorithm for various test cases. These tests were run on an *SGI Indigo II* with an R4400 CPU, and double precision floating point numbers were used for the calculations.

The set of polyhedral objects that have volume integrals which are commonly tabulated or easy to compute by hand is rather limited. We ran our algorithm on two such objects: an axes-aligned cube, with 20 units on a side and centered at the origin; and a tetrahedron defined by the convex hull of the origin, and the vectors $5\hat{\mathbf{i}}$, $4\hat{\mathbf{j}}$, and $3\hat{\mathbf{k}}$. The theoretical values for these objects are shown in Table 1. For these two examples, all values computed by the algorithm were correct to at least 15 significant figures. The total time required to compute all ten volume integrals was 64 μs for the tetrahedron, and 110 μs for the cube.

object	T_1	T_x	T_y	T_z	T_{x^2}	T_{y^2}	T_{z^2}	T_{xy}	T_{yz}	T_{zx}
cube	8000	0	0	0	2.67×10^5	2.67×10^5	2.67×10^5	0	0	0
tetrahedron	10	12.5	10	7.5	25	16	9	10	6	7.5

Table 1. Theoretical values of volume integrals for simple test polyhedra.

For a more interesting test, the algorithm was applied to several polyhedral approximations of a unit radius sphere, centered at the origin. In this case there are two sources of error: numerical errors from the algorithm, and approximation errors inherent in the geometric model, which is not a true sphere. These latter errors should not be attributed to the algorithm itself. For a perfect unit sphere, the integrals $T_x, T_y, T_z, T_{xy}, T_{yz}$, and T_{zx} should vanish, while $T_1 = \frac{4}{3}\pi$ and $T_{x^2} = T_{y^2} = T_{z^2} = \frac{4}{15}\pi$. We applied our algorithm to a series of successive approximations to the sphere, beginning with an icosahedron, and obtaining each finer approximation by projecting the midpoint of each polyhedral edge onto the unit sphere, and taking a convex hull. The computed values of a representative set of volume integrals for each polyhedron are shown in Table 2.

approx.	verts	edges	faces	T_1	T_x	T_{x^2}	T_{yz}	time
1	12	30	20	2.536	-2.8×10^{-17}	0.3670	-3.1×10^{-17}	500 μs
2	42	120	80	3.659	$+1.4 \times 10^{-16}$	0.6692	$+1.5 \times 10^{-17}$	1.2 ms
3	162	480	320	4.047	-3.2×10^{-16}	0.7911	-6.1×10^{-18}	4.9 ms
4	642	1920	1280	4.153	$+3.0 \times 10^{-16}$	0.8258	$+7.8 \times 10^{-18}$	21 ms
5	2562	7680	5120	4.180	-3.8×10^{-17}	0.8347	$+2.1 \times 10^{-17}$	82 ms
6	10242	30720	20480	4.187	$+5.6 \times 10^{-16}$	0.8370	$+6.4 \times 10^{-18}$	350 ms
theoretical values for sphere				4.189	0.0	0.8378	0.0	–

Table 2. Data for successive approximations of a unit sphere.

Without numerical error, the integrals T_x and T_{yz} would vanish for all six polyhedral approximations of the sphere, due to symmetry. From Table 2, the absolute values of these computed values are all less than 10^{-15}. The theoretical values in the table correspond to the sphere which circumscribes the polyhedra. For each polyhedron, we have also determined corresponding values for the inscribed sphere, and verified that the computed values for T_1 and T_{x^2} for the polyhedron lie between the bounding values for these two spheres. For approximation 6, the difference in values for the inscribed and circumscribed sphere is 2.8×10^{-3} for T_1 and 9.5×10^{-4} for T_{x^2}. These values are upper bounds on the numerical errors of the algorithm. Note that the deviations between theoretical and computed values for T_1 and T_{x^2} are reduced as the complexity of the polyhedron increases, while numerical error from the algorithm should increase with complexity. In light of the very small errors incurred in the computation of T_x and T_{yz}, the deviations between the computed and theoretical values of T_1 and T_{x^2} are almost certainly due mainly to the polyhedral approximation rather than to numerical errors.

The execution times shown in Table 2 are the total times for computing all ten volume integrals for each polyhedron. The $O(n)$ nature of the algorithm is evident: from approximation 2 on, the time ratios between successive refinements very closely follow the 4:1 ratio in the number of faces. The algorithm is also very fast: all ten integrals are computed for a polyhedron with over 20,000 faces in only 350 ms.

10.   Available Code

ANSI C source code for the algorithm described in this paper, and detailed in Figures 4–6, is publicly available from

http://www.acm.org/jgt .

Also included is an interface to build up polyhedra (using a simple data structure) from ASCII specifications; examples are provided. The code is public domain, and may be used as is or in modified form.

CompVolumeIntegrals($\mathcal{V}$)

$$T_1, T_x, T_y, T_z, T_{x^2}, T_{y^2}, T_{z^2}, T_{xy}, T_{yz}, T_{zx} \leftarrow 0$$

```
for each face F on the boundary of V
    choose α-β-γ as a right-handed permutation of x-y-z
        that maximizes |n̂γ|
    compFaceIntegrals(F)
```
$\qquad$ if $(\alpha = x)$ $T_1 \leftarrow T_1 + \hat{n}_\alpha F_\alpha$
$\qquad$ else if $(\beta = x)$ $T_1 \leftarrow T_1 + \hat{n}_\beta F_\beta$
$\qquad$ else $T_1 \leftarrow T_1 + \hat{n}_\gamma F_\gamma$
$\qquad$ $T_\alpha \leftarrow T_\alpha + \hat{n}_\alpha F_{\alpha^2}$
$\qquad$ $T_\beta \leftarrow T_\beta + \hat{n}_\beta F_{\beta^2}$
$\qquad$ $T_\gamma \leftarrow T_\gamma + \hat{n}_\gamma F_{\gamma^2}$
$\qquad$ $T_{\alpha^2} \leftarrow T_{\alpha^2} + \hat{n}_\alpha F_{\alpha^3}$
$\qquad$ $T_{\beta^2} \leftarrow T_{\beta^2} + \hat{n}_\beta F_{\beta^3}$
$\qquad$ $T_{\gamma^2} \leftarrow T_{\gamma^2} + \hat{n}_\gamma F_{\gamma^3}$
$\qquad$ $T_{\alpha\beta} \leftarrow T_{\alpha\beta} + \hat{n}_\alpha F_{\alpha^2\beta}$
$\qquad$ $T_{\beta\gamma} \leftarrow T_{\beta\gamma} + \hat{n}_\beta F_{\beta^2\gamma}$
$\qquad$ $T_{\gamma\alpha} \leftarrow T_{\gamma\alpha} + \hat{n}_\gamma F_{\gamma^2\alpha}$

$$(T_x, T_y, T_z) \leftarrow (T_x, T_y, T_z) \ / \ 2$$
$$(T_{x^2}, T_{y^2}, T_{z^2}) \leftarrow (T_{x^2}, T_{y^2}, T_{z^2}) \ / \ 3$$
$$(T_{xy}, T_{yz}, T_{zx}) \leftarrow (T_{xy}, T_{yz}, T_{zx}) \ / \ 2$$

Figure 4. CompVolumeIntegrals($\mathcal{V}$). Compute the required volume integrals for a polyhedron. See Equations (29)–(38).

CompFaceIntegrals($\mathcal{F}$)

  computeProjectionIntegrals($\mathcal{F}$)
  $w \leftarrow -\hat{\mathbf{n}} \cdot \mathbf{p}$ for some point $\mathbf{p}$ on $\mathcal{F}$
  $k_1 \leftarrow \hat{n}_\gamma^{-1}; \quad k_2 \leftarrow k_1 * k_1; \quad k_3 \leftarrow k_2 * k_1; \quad k_4 \leftarrow k_3 * k_1$

  $F_\alpha \leftarrow k_1 \pi_\alpha$
  $F_\beta \leftarrow k_1 \pi_\beta$
  $F_\gamma \leftarrow -k_2(\hat{n}_\alpha \pi_\alpha + \hat{n}_\beta \pi_\beta + w\pi_1)$

  $F_{\alpha^2} \leftarrow k_1 \pi_{\alpha^2}$
  $F_{\beta^2} \leftarrow k_1 \pi_{\beta^2}$
  $F_{\gamma^2} \leftarrow k_3(\hat{n}_\alpha^2 \pi_{\alpha^2} + 2\hat{n}_\alpha \hat{n}_\beta \pi_{\alpha\beta} + \hat{n}_\beta^2 \pi_{\beta^2} + 2\hat{n}_\alpha w\pi_\alpha + 2\hat{n}_\beta w\pi_\beta + w^2\pi_1)$

  $F_{\alpha^3} \leftarrow k_1 \pi_{\alpha^3}$
  $F_{\beta^3} \leftarrow k_1 \pi_{\beta^3}$
  $F_{\gamma^3} \leftarrow -k_4(\hat{n}_\alpha^3 \pi_{\alpha^3} + 3\hat{n}_\alpha^2 \hat{n}_\beta \pi_{\alpha^2\beta} + 3\hat{n}_\alpha \hat{n}_\beta^2 \pi_{\alpha\beta^2} + \hat{n}_\beta^3 \pi_{\beta^3} +$
  $\qquad 3\hat{n}_\alpha^2 w\pi_{\alpha^2} + 6\hat{n}_\alpha \hat{n}_\beta w\pi_{\alpha\beta} + 3\hat{n}_\beta^2 w\pi_{\beta^2} + 3\hat{n}_\alpha w^2\pi_\alpha + 3\hat{n}_\beta w^2\pi_\beta + w^3\pi_1)$

  $F_{\alpha^2\beta} \leftarrow k_1 \pi_{\alpha^2\beta}$
  $F_{\beta^2\gamma} \leftarrow -k_2(\hat{n}_\alpha \pi_{\alpha\beta^2} + \hat{n}_\beta \pi_{\beta^3} + w\pi_{\beta^2})$
  $F_{\gamma^2\alpha} \leftarrow k_3(\hat{n}_\alpha^2 \pi_{\alpha^3} + 2\hat{n}_\alpha \hat{n}_\beta \pi_{\alpha^2\beta} + \hat{n}_\beta^2 \pi_{\alpha\beta^2} + 2\hat{n}_\alpha w\pi_{\alpha^2} + 2\hat{n}_\beta w\pi_{\alpha\beta} + w^2\pi_\alpha)$

Figure 5. CompFaceIntegrals($\mathcal{F}$). Compute the required surface integrals over a polyhedral face. See Equations (57)–(68).

CompProjectionIntegrals($\mathcal{F}$)

$\pi_1, \pi_\alpha, \pi_b, \pi_{\alpha^2}, \pi_{\alpha\beta}, \pi_{\beta^2}, \pi_{\alpha^3}, \pi_{\alpha^2\beta}, \pi_{\alpha\beta^2}, \pi_{\beta^3} \leftarrow 0$

for each edge $\mathcal{E}$ in CCW order around $\mathcal{F}$
$\quad \alpha_0 \leftarrow \alpha$-coordinate of start point of $\mathcal{E}$
$\quad \beta_0 \leftarrow \beta$-coordinate of start point of $\mathcal{E}$
$\quad \alpha_1 \leftarrow \alpha$-coordinate of end point of $\mathcal{E}$
$\quad \beta_1 \leftarrow \beta$-coordinate of end point of $\mathcal{E}$
$\quad \Delta\alpha \leftarrow \alpha_1 - \alpha_0$
$\quad \Delta\beta \leftarrow \beta_1 - \beta_0$
$\quad \alpha_0^2 \leftarrow \alpha_0 * \alpha_0 \ ; \ \alpha_0^3 \leftarrow \alpha_0^2 * \alpha_0 \ ; \ \alpha_0^4 \leftarrow \alpha_0^3 * \alpha_0$
$\quad \beta_0^2 \leftarrow \beta_0 * \beta_0 \ ; \ \beta_0^3 \leftarrow \beta_0^2 * \beta_0 \ ; \ \beta_0^4 \leftarrow \beta_0^3 * \beta_0$
$\quad \alpha_1^2 \leftarrow \alpha_1 * \alpha_1 \ ; \ \alpha_1^3 \leftarrow \alpha_1^2 * \alpha_1$
$\quad \beta_1^2 \leftarrow \beta_1 * \beta_1 \ ; \ \beta_1^3 \leftarrow \beta_1^2 * \beta_1$

$\quad C_1 \leftarrow \alpha_1 + \alpha_0$
$\quad C_\alpha \leftarrow \alpha_1 C_1 + \alpha_0^2 \ ; \ C_{\alpha^2} \leftarrow \alpha_1 C_\alpha + \alpha_0^3 \ ; \ C_{\alpha^3} \leftarrow \alpha_1 C_{\alpha^2} + \alpha_0^4$
$\quad C_\beta \leftarrow \beta_1^2 + \beta_1 \beta_0 + \beta_0^2 \ ; \ C_{\beta^2} \leftarrow \beta_1 C_\beta + \beta_0^3 \ ; \ C_{\beta^3} \leftarrow \beta_1 C_{\beta^2} + \beta_0^4$
$\quad C_{\alpha\beta} \leftarrow 3\alpha_1 + 2\alpha_1\alpha_0 + \alpha_0^2 \ ; \ K_{\alpha\beta} \leftarrow \alpha_1 + 2\alpha_1\alpha_0 + 3\alpha_0^2$
$\quad C_{\alpha^2\beta} \leftarrow \alpha_0 C_{\alpha\beta} + 4\alpha_1^3 \ ; \ K_{\alpha^2\beta} \leftarrow \alpha_1 K_{\alpha\beta} + 4\alpha_0^3$
$\quad C_{\alpha\beta^2} \leftarrow 4\beta_1^3 + 3\beta_1^2\beta_0 + 2\beta_1\beta_0^2 + \beta_0^3 \ ;$
$\quad K_{\alpha\beta^2} \leftarrow \beta_1^3 + 2\beta_1^2\beta_0 + 3\beta_1\beta_0^2 + 4\beta_0^3$

$\quad \pi_1 \leftarrow \pi_1 + \Delta\beta C_1$
$\quad \pi_\alpha \leftarrow \pi_\alpha + \Delta\beta C_\alpha \ ; \ \pi_{\alpha^2} \leftarrow \pi_{\alpha^2} + \Delta\beta C_{\alpha^2} \ ; \ \pi_{\alpha^3} \leftarrow \pi_{\alpha^3} + \Delta\beta C_{\alpha^3}$
$\quad \pi_\beta \leftarrow \pi_\beta + \Delta\alpha C_\beta \ ; \ \pi_{\beta^2} \leftarrow \pi_{\beta^2} + \Delta\alpha C_{\beta^2} \ ; \ \pi_{\beta^3} \leftarrow \pi_{\beta^3} + \Delta\alpha C_{\beta^3}$
$\quad \pi_{\alpha\beta} \leftarrow \pi_{\alpha\beta} + \Delta\beta(\beta_1 C_{\alpha\beta} + \beta_0 K_{\alpha\beta})$
$\quad \pi_{\alpha^2\beta} \leftarrow \pi_{\alpha^2\beta} + \Delta\beta(\beta_1 C_{\alpha^2\beta} + \beta_0 K_{\alpha^2\beta})$
$\quad \pi_{\alpha\beta^2} \leftarrow \pi_{\alpha\beta^2} + \Delta\alpha(\alpha_1 C_{\alpha\beta^2} + \alpha_0 K_{\alpha\beta^2})$

$\pi_1 \leftarrow \pi_1 / 2$
$\pi_\alpha \leftarrow \pi_\alpha / 6 \ ; \ \pi_{\alpha^2} \leftarrow \pi_{\alpha^2} / 12 \ ; \ \pi_{\alpha^3} \leftarrow \pi_{\alpha^3} / 20$
$\pi_\beta \leftarrow -\pi_\beta / 6 \ ; \ \pi_{\beta^2} \leftarrow -\pi_{\beta^2} / 12 \ ; \ \pi_{\beta^3} \leftarrow -\pi_{\beta^3} / 20$
$\pi_{\alpha\beta} \leftarrow \pi_{\alpha\beta} / 24$
$\pi_{\alpha^2\beta} \leftarrow \pi_{\alpha^2\beta} / 60$
$\pi_{\alpha\beta^2} \leftarrow -\pi_{\alpha\beta^2} / 60$

Figure 6. CompProjectionIntegrals($\mathcal{F}$). Compute the required integrals over a face projection. See Equations (89)–(98).

Vol. 1, No. 2: 31–50

Acknowledgments. We thank Aristides Requicha for a valuable literature survey on this topic, and David Baraff for useful comments on the initial draft of this paper. We especially thank John Hughes for his detailed review and many suggestions for improving the paper.

References

[Edwards, Penney 86] C. H. Edwards, Jr. and David E. Penney. *Calculus and Analytic Geometry*. Englewood Cliffs, NJ: Prentice-Hall, second edition, 1986.

[Farin 90] Gerald Farin. *Curves and Surfaces for Computer Aided Geometric Design*. San Diego: Academic Press, second edition, 1990.

[Farouki, Rajan 88] R. T. Farouki and V. T. Rajan. "Algorithms for Polynomials in Bernstein Form." *Computer Aided Geometric Design*, 5(1):1–26(June 1988).

[Greenwood 88] Donald T. Greenwood. *Principles of Dynamics*. Englewood Cliffs, NJ: Prentice-Hall, second edition, 1988.

[Lee, Requicha 82] Yong Tsui Lee and Aristides A. G. Requicha. "Algorithms for Computing the Volume and Other Integral Properties of Solids. I. Known Methods and Open Issues." *Communications of the ACM*, 25(9):635–41 (September 1982).

[Lien, Kajiya 84] Sheue-ling Lien and James T. Kajiya. "A Symbolic Method for Calculating the Integral Properties of Arbitrary Nonconvex Polyhedra." *IEEE Computer Graphics and Applications*, 4(10):35–41(October 1984).

[Meriam, Kraige 86] J. L. Meriam and L. G. Kraige. *Engineering Mechanics Volume 2: Dynamics*. New York: John Wiley & Sons, 1986.

[Press et al. 92] William H. Press, Saul A. Teukolsky, William T. Vetterling, and Brian R. Flannery. *Numerical Recipes in C: The Art of Scientific Computing*. Cambridge, UK: Cambridge University Press, second edition, 1992.

[Stein 82] Sherman K. Stein. *Calculus and Analytic Geometry*. New York: McGraw-Hill, third edition, 1982.

[Wylie, Barrett 82] C. Ray Wylie and Louis C. Barrett. *Advanced Engineering Mathematics*. New York: McGraw-Hill, fifth edition, 1982.

Web Information:

ANSI C source code for the algorithm described is available at
http://www.acm.org/jgt

Brian Mirtich, University of California at Berkeley, Berkeley, CA
(mirtich@cahors.cs.berkeley.edu)

Editor: John F. Hughes

Received September 6, 1995; accepted October 2, 1995

Current Contact Information:

Brian Mirtich, Cognex Corporation, One Vision Drive, Natick, MA 01760
(brian.mirtich@cognex.com)

Vol. 2, No. 4: 1–13

Efficient Collision Detection of Complex Deformable Models using AABB Trees

Gino van den Bergen
Eindhoven University of Technology

Abstract. We present a scheme for exact collision detection between complex models undergoing rigid motion and deformation. The scheme relies on a hierarchical model representation using axis-aligned bounding boxes (AABBs). Recent work has shown that AABB trees are slower than oriented bounding box (OBB) trees for performing overlap tests. In this paper, we describe a way to speed up overlap tests between AABBs, such that for collision detection of rigid models, the difference in performance between the two representations is greatly reduced. Furthermore, we show how to update an AABB tree quickly as a model is deformed. We thus find AABB trees to be the method of choice for collision detection of complex models undergoing deformation. In fact, because they are not much slower to test, are faster to build, and use less storage than OBB trees, AABB trees might be a reasonable choice for rigid models as well.

1. Introduction

Hierarchies of bounding volumes provide a fast way to perform exact collision detection between complex models. Examples of volume types that are used for this purpose include spheres [Palmer, Grimsdale 95], [Hubbard 96], oriented bounding boxes (OBBs) [Gottschalk et al. 96], and discrete-orientation polytopes (DOPs) [Klosowksi et al. 98], [Zachmann 98]. In this paper, we present a collision-detection scheme that relies on a hierarchical model representation using axis-aligned bounding boxes (AABBs). In the AABB trees as

131

we use them, the boxes are aligned to the axes of the model's local coordinate system, thus all the boxes in a tree have the same orientation.

In recent work [Gottschalk et al. 96], AABB trees have been shown to yield worse performance than OBB trees for overlap tests of rigid models. In this paper, however, we present a way to speed up overlap testing between relatively oriented boxes of a pair of AABB trees. This results in AABB tree performance that is close to the OBB tree's performance for collision detection of rigid models.

Furthermore, we show how to update an AABB tree as quickly as a model is deformed. Updating an AABB tree after a deformation is considerably faster than rebuilding the tree, and results in a tight-fitting hierarchy of boxes for most types of deformations. Updating an OBB tree is significantly more complex.

In comparison to a previous algorithm for deformable models [Smith et al. 95], the algorithm presented here is expected to perform better for deformable models that are placed in close proximity. For these cases, both algorithms show a time complexity that is roughly linear in the number of primitives. However, our approach has a smaller constant (asymptotically 48 arithmetic operations per triangle for triangle meshes). Moreover, our algorithm is better suited for collision detection between a mix of rigid and deformable models, since it is linear in the number of primitives of the deformable models only.

The C++ source code for the scheme presented here is released as part of the Software Library for Interference Detection (SOLID) version 2.0.

2.   Building an AABB Tree

The AABB tree that we consider is, as is the OBB tree described by Gottschalk [Gottschalk et al. 96], a binary tree. The two structures differ with respect to the freedom of placement of the bounding boxes: AABBs are aligned to the axes of the model's local coordinate system, whereas OBBs can be arbitrarily oriented. The added freedom of an OBB is gained at a considerable cost in storage space. An OBB is represented using 15 scalars (9 scalars for a 3×3 matrix representing the orientation, 3 scalars for position, and 3 for extent), whereas an AABB requires only 6 scalars (for position and extent). Hence, an AABB tree of a model requires roughly half as much storage space as an OBB tree of the same model.

An AABB tree is constructed top-down, by recursive subdivision. In each recursion step, the smallest AABB of the set of primitives is computed, and the set is split by ordering the primitives according to a well-chosen partitioning plane. This process continues until each subset contains one element. Thus, an AABB tree for a set of n primitives has n leaves and $n - 1$ internal nodes.

In order to get a 'fat' subdivision (i.e., cube-like rather than oblong), the partitioning plane is chosen orthogonal to the longest axis of the AABB. In

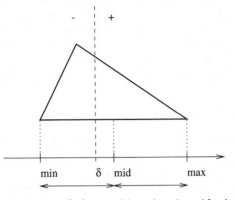

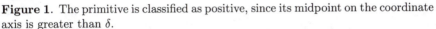

Figure 1. The primitive is classified as positive, since its midpoint on the coordinate axis is greater than δ.

general, fat AABBs yield better performance in intersection testing, since under the assumption that the boxes in a tree mutually overlap as little as possible, a given query box can overlap fewer fat boxes than thin boxes.

We position the partitioning plane along the longest axis, by choosing δ, the coordinate on the longest axis where the partitioning plane intersects the axis. We then split the set of primitives into a negative and positive subset corresponding to the respective halfspaces of the plane. A primitive is classified as positive if the midpoint of its projection onto the axis is greater than δ, and negative otherwise. Figure 1 shows a primitive that straddles the partitioning plane depicted by a dashed line. This primitive is classified as positive. It can be seen that by using this subdivision method, the degree of overlap between the AABBs of the two subsets is kept small.

For choosing the partitioning coordinate δ, we tried several heuristics. Our experiments with AABB trees for a number of polygonal models showed us that, in general, the best performance is achieved by simply choosing the δ to be the median of the AABB, thus splitting the box in two equal halves. Using this heuristic, it may take $O(n^2)$ time in the worst case to build an AABB tree for n primitives, however, in the usual case where the primitives are distributed more or less uniformly over the box, building an AABB tree takes only $O(n \log n)$ time. Other heuristics we have tried that didn't perform as well are: (a) subdividing the set of primitives in two sets of equal size, thus building an optimally-balanced tree, and (b) building a halfbalanced tree, i.e., the larger subset is at most twice as large as the smaller one, and the overlap of the subsets' AABBs projected onto the longest axis is minimized.

Occasionally, it may occur that all primitives are classified to the same side of the plane. This will happen most frequently when the set of primitives contains only a few elements. In this case, we simply split the set in two subsets of (almost) equal size, disregarding the geometric location of the primitives.

Building an AABB tree of a given model is faster than building an OBB tree for that model, since the estimation of the best orientation of an OBB for a given set of primitives requires additional computations. We found that building an OBB tree takes about three times as much time as building an AABB tree, as shown in Section 5.

3. Intersection Testing

An intersection test between two models is done by recursively testing pairs of nodes. For each visited pair of nodes, the AABBs are tested for overlap. Only the nodes for which the AABBs overlap are further traversed. If both nodes are leaves, then the primitives are tested for intersection and the result of the test is returned. Otherwise, if one of the nodes is a leaf and the other an internal node, then the leaf node is tested for intersection with each of the children of the internal node. Finally, if both nodes are internal nodes, then the node with smaller volume is tested for intersection with the children of the node with the larger volume. The latter heuristic choice of unfolding the node with the largest volume results in the largest reduction of total volume size in the following AABB tests thus the lowest probability of the following tested boxes overlapping.

Since the local coordinate systems of a pair of models may be arbitrarily oriented, we need an overlap test for relatively oriented boxes. A fast overlap test for oriented boxes is presented by Gottschalk [Gottschalk et al. 96]. We will refer to this test as the *separating axes test* (SAT). A separating axis of two boxes is an axis for which the projections of the boxes onto the axis do not overlap. The existence of a separating axis for a pair of boxes sufficiently classifies the boxes as disjoint. It can be shown that for any disjoint pair of convex three-dimensional polytopes, a separating axis can be found that is either orthogonal to a facet of one of the polytopes, or orthogonal to an edge from each polytope [Gottschalk 96b]. This results in 15 potential separating axes that need to be tested for a pair of oriented boxes (3 facet orientations per box plus 9 pairwise combinations of edge directions). The SAT exits as soon as a separating axis is found. If none of the 15 axes separate the boxes, then the boxes overlap.

We refer to the Gottschalk paper [Gottschalk et al. 96] for details on how the SAT is implemented such that it uses the least number of operations. For the following discussion, it is important to note that this implementation requires the relative orientation represented by a 3×3 matrix and its absolute value, i.e., the matrix of absolute values of matrix elements, be computed before performing the 15 axes tests.

In general, testing two AABB trees for intersection requires more box-overlap tests than testing two OBB trees of the same models, since the smallest AABB of a set of primitives is usually larger than the smallest OBB. How-

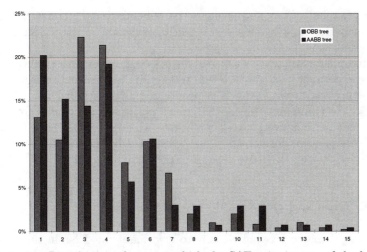

Figure 2. Distribution of axes on which the SAT exits in case of the boxes being disjoint. Axes 1 to 6 correspond to the facet orientations of the boxes, and axes 7 to 15 correspond to the combinations of edge directions.

ever, since each tested pair of boxes of two OBB trees normally has a different relative orientation, the matrix operations for computing this orientation and its absolute value are repeated for each tested pair of boxes. For AABB trees, however, the relative orientation is the same for each tested pair of boxes, and thus needs to be computed only once. Therefore, the performance of an AABB tree might not be as bad as we would expect. The empirical results in Section 5. show that, by exploiting this feature, intersection testing using AABB trees usually takes only 50% longer than using OBB trees in cases where there is a lot of overlap among the models.

For both tree types, the most time-consuming operation in the intersection test is the SAT, so let us see if there is room for improvement. We found that, in the case where the boxes are disjoint, the probability of the SAT exiting on an axis corresponding to a pair of edge directions is about 15%. Figure 2 shows a distribution of the separating axes on which the SAT exits for tests with a high probability of the models intersecting. Moreover, for both the OBB and the AABB tree, we found that about 60% of all box-overlap tests resulted in a positive result. Thus, if we remove from the SAT the nine axis tests corresponding to the edge directions, we will get an incorrect result only 6% (40% of 15%) of the time.

Since the box-overlap test is used for quick rejection of subsets of primitives, exact determination of a box-overlap is not necessary. Using a box-overlap test, that returns more overlaps than there actually are, results in more nodes being visited, and thus more box-overlap and primitive intersection tests. Examining fewer axes in the SAT reduces the cost of a box-overlap test, but increases the number of box and primitive pairs being tested. Thus, there is a trade-off of pertest cost against number of tests, when we use a SAT that examines fewer axes.

In order to find out whether this trade-off works in favor of performance, we repeated the experiment using a SAT that tests only the six facet orientations. We refer to this test as *SAT lite*. The results of this experiment are shown in Section 5. We found that the AABB tree's performance benefits from a cheaper but sloppier box-overlap test in all cases, whereas the OBB tree shows hardly any change in performance. This is explained by the higher cost of a box-overlap test for the OBB tree due to extra matrix operations.

4.   AABB Trees and Deformable Models

AABB trees lend themselves quite easily to deformable models. In this context, a deformable model is a set of primitives in which the placements and shapes of the primitives within the model's local coordinate system change over time. A typical example of a deformable model is a triangle mesh in which the local coordinates of the vertices are time-dependent.

Instead of rebuilding the tree after a deformation, it is usually a lot faster to refit the boxes in the tree. The following property of AABBs allows an AABB tree to be refitted efficiently in a bottom-up manner. Let S be a set of primitives and S^+, S^- subsets of S such that $S^+ \cup S^- = S$, and let B^+ and B^- be the smallest AABBs of respectively S^+ and S^-, and B, the smallest AABB enclosing $B^+ \cup B^-$. Then, B is also the smallest AABB of S.

This property is illustrated in Figure 3. Of all bounding volume types we have seen so far, AABBs share this property only with DOPs.

This property of AABBs yields a straightforward method for refitting a hierarchy of AABBs after a deformation. First the bounding boxes of the leaves are recomputed, after which each parent box is recomputed using the boxes of its children in a strict bottom-up order. This operation may be implemented as a postorder tree traversal, i.e., for each internal node, the children are visited first, after which the bounding box is recomputed. However, in order to avoid the overhead of recursive function calls, we implement it differently.

In our implementation, the leaves and the internal nodes of an AABB tree are allocated as arrays of nodes. We are able to do this, since the number of primitives in the model is static. Furthermore, the tree is built such that each internal child node's index number in the array is greater than its parent's index number. In this way, the internal nodes are refitted properly by iterating over the array of internal nodes in reverse order. Since refitting an AABB takes constant time for both internal nodes and leaves, an AABB tree is refitted in time linear in the number of nodes. Refitting an AABB tree of a triangle mesh takes less than 48 arithmetic operations per triangle. Experiments have

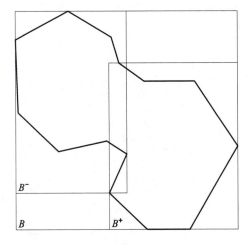

Figure 3. The smallest AABB of a set of primitives encloses the smallest AABBs of the subsets in a partition of the set.

shown that for models composed of over 6,000 triangles, refitting an AABB tree is about ten times as fast as rebuilding it.

There is, however, a drawback to this method of refitting. Due to relative position changes of primitives in the model after a deformation, the boxes in a refitted tree may have a higher degree of overlap than the boxes in a rebuilt tree. Figure 4 illustrates this effect for the model in Figure 3. A higher degree of overlap of boxes in the tree results in more nodes being visited during an intersection test, and thus, worse performance for intersection testing.

We observe a higher degree of overlap among the boxes in a refitted tree mostly for radical deformations such as excessive twists, features blown out of proportion, or extreme forms of self-intersection. However, for deformations that keep the adjacency relation of triangles in a mesh intact (i.e., the mesh is not torn up), we found no significant performance deterioration for inter-section testing, even for the more severe deformations. This is due to the fact that the degree of overlap increases mostly for the boxes that are maintained high in the tree, whereas most of the box-overlap tests are performed on boxes that are maintained close to the leaves.

5.   Performance

The total cost of testing a pair of models represented by bounding volume hierarchies is expressed in the following cost function [Weghorst et al. 94], [Gottschalk et al. 96]:

$$T_{\text{total}} = N_b * C_b + N_p * C_p,$$

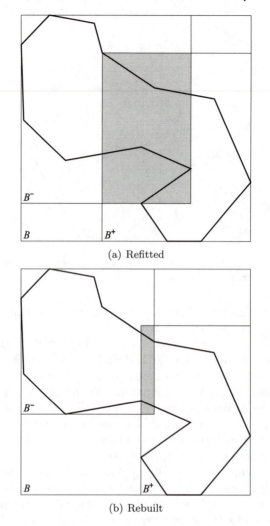

(a) Refitted

(b) Rebuilt

Figure 4. Refitting vs. rebuilding the model in Figure 3 after a deformation.

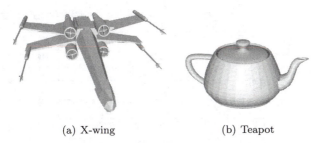

(a) X-wing   (b) Teapot

Figure 5. Two models used in our experiments.

where

T_{total}	is the total cost of testing a pair of models for intersection,
N_b	is the number of bounding volume pairs tested for overlap,
C_b	is the cost of testing a pair of bounding volumes for overlap,
N_p	is the number of primitive pairs tested for intersection, and
C_p	is the cost of testing a pair of primitives for intersection.

The parameters in the cost function that are affected by the choice of bounding volume are N_b, N_p, and C_b. A tight-fitting bounding volume type, such as an OBB, results in a low N_b and N_p but a relatively high C_b, whereas an AABB will result in more tests being performed and a lower value of C_b.

In order to compare the performances of the AABB tree and the OBB tree, we have conducted an experiment in which a pair of models were placed randomly in a bounded space and tested for intersection. The random orientations of the models were generated using the method described by Shoemake [Shoemake 92]. The models were positioned by placing the origin of each model's local coordinate system randomly inside a cube. The probability of an intersection is tuned by changing the size of the cube. For all tests, the probability was set to approximately 60%.

For this experiment, we used Gottschalk's RAPID package [Gottschalk 96a] for the OBB tree tests. For the AABB tree tests, we used a modified RAPID, in which we removed the unnecessary matrix operations. We experimented with three models: a torus composed of 5,000 triangles, a slender X-wing spacecraft composed of 6,084 triangles, and the archetypical teapot composed of 3,752 triangles, as shown in Figure 5. Each test performed 100,000 random placements and intersection tests, resulting in approximately 60,000 collisions for all tested models. Table 1 shows the results of the tests for both the OBB tree and the AABB tree. The tests were performed on a Sun UltraSPARC-I (167MHz), compiled using the GNU compiler with '-O2' optimization.

OBB tree							
Model	N_b	C_b	T_b	N_p	C_p	T_p	T_{total}
Torus	10178961	4.9	49.7	197314	15	2.9	52.6
X-wing	48890612	4.6	223.8	975217	10	10.2	234.0
Teapot	12025710	4.8	57.6	186329	14	2.7	60.3

AABB tree							
Model	N_b	C_b	T_b	N_p	C_p	T_p	T_{total}
Torus	32913297	3.7	122.3	3996806	7.2	28.7	151.0
X-wing	92376250	3.1	288.8	8601433	7.1	61.3	350.1
Teapot	25810569	3.3	84.8	1874830	7.4	13.9	98.7

Table 1. Performance of the AABB tree vs. the OBB tree, both using the SAT. N_b and N_p are respectively the total number box and triangle intersection tests, C_b and C_p the per-test times in microseconds for respectively the box and triangle intersection test, $T_b = N_b * C_b$ is the total time in seconds spent testing for box intersections, $T_p = N_p * C_p$ is the total time used for triangle intersection tests, and finally T_{total} is the total time in seconds for performing 100K intersection tests.

An AABB tree requires approximately twice as many box-intersection tests as an OBB tree; however, the time used for intersection testing is in most cases only 50% longer for AABB trees. The exception here is the torus model, for which the AABB tree takes almost three times as long as the OBB tree. Apparently, the OBB tree excels in fitting models that have a smooth surface composed of uniformly distributed primitives. Furthermore, we observe that, due to its tighter fit, the OBB tree requires much fewer triangle-intersection tests (less than two triangle-intersection tests per placement, for the torus and the teapot).

We repeated the experiment using a separating axes test that tests only the axes corresponding to the six facet orientations, referred to as *SAT lite*. The results of this experiment are shown in Table 2. We see a performance increase of about 15% on average for the AABB tree, whereas the change in performance for the OBB tree is only marginal.

We also ran some tests in order to see how the time used for refitting an AABB tree for a deformable model compares to the intersection-testing time. We found that on our testing platform, refitting a triangle mesh composed of a large number of triangles ($> 1,000$) takes 2.9 microseconds per triangle. For instance, for a pair of models composed of 5,000 triangles each, refitting takes 29 milliseconds, which is more than 10 times the amount of time it takes to test the models for intersection. Hence, refitting is likely to become the bottleneck if many of the models in a simulated environment are deformed and refitted in each frame. However, for environments with many moving

OBB tree							
Model	N_b	C_b	T_b	N_p	C_p	T_p	T_{total}
Torus	13116295	3.7	47.9	371345	12	4.4	52.3
X-wing	65041340	3.4	221.4	2451543	9.3	22.9	244.3
Teapot	14404588	3.5	50.8	279987	13	3.5	54.3
AABB tree							
Model	N_b	C_b	T_b	N_p	C_p	T_p	T_{total}
Torus	40238149	2.4	96.1	5222836	7.4	38.4	134.5
X-wing	121462120	1.9	236.7	13066095	7.0	91.3	328.0
Teapot	30127623	2.1	62.5	2214671	7.0	15.6	78.1

Table 2. Performance of AABB tree vs. OBB tree, both using the SAT lite

models, in which only a few are deformed in each frame, refitting will not take much more time in total than intersection testing.

We conclude with a comparison of the performance of the AABB tree vs. the OBB tree for deformable models. Table 3 presents an overview of the times we found for operations on the two tree types. We see that for deformable models, the OBB's faster intersection test is not easily going to make up for the high cost of rebuilding the OBB trees, even if only a few of the models are deformed. For these cases, AABB trees, which are refitted in less than 5% of the time it takes to rebuild an OBB tree, will yield better performance, and are therefore the preferred method for collision detection of deformable models.

6. Implementation Notes

In SOLID 2.0, AABB trees are used both for rigid and deformable models. In order to comply with the structures and motions specified in VRML [Bell et al. 97], SOLID allows, besides translations and rotations, also nonuniform scalings on models. Note that a nonuniform scaling is not considered a deformation, and hence does not require refitting. However, in order to be able to

Operation	Torus	X-wing	Teapot
(Re)build an OBB tree	0.35s	0.46s	0.27s
Build an AABB tree	0.11s	0.18s	0.08s
Refit an AABB tree	15ms	18ms	11ms
Test a pair of OBB trees	0.5ms	2.3ms	0.6ms
Test a pair of AABB trees	1.3ms	3.3ms	0.8ms

Table 3. Comparing the times for a number of operations

use nonuniformly scaled models, some changes in the AABB overlap test are needed.

Let $\mathbf{T}(\mathbf{x}) = \mathbf{Bx} + \mathbf{c}$ be the relative transformation from a model's local coordinate system to the local coordinate system of another model, where $\mathbf{B}$ is a 3×3 matrix, representing the orientation and scaling, and $\mathbf{c}$ is the vector representing the translation. For nonuniformly scaled models, we cannot rely on the matrix $\mathbf{B}$ being orthogonal, i.e., $\mathbf{B}^{-1} = \mathbf{B}^{\mathrm{T}}$. However, for SAT lite, both $\mathbf{B}$ and $\mathbf{B}^{-1}$ and their respective absolute values are needed. Hence, in our implementation, we compute these four matrices for each intersection test of a pair of models, and use them for each tested pair of boxes. The added cost of allowing nonuniformly scaled models is negligible, since $\mathbf{B}^{-1}$ and its absolute value are computed only once for each tested pair of models.

Finally, it is worth mentioning that for AABB trees, a larger percentage of time is used for primitive intersection tests than for OBB trees (28% vs. 5%). In this respect, it might be a good idea to use the triangle-intersection test presented by Möller [Möller 97], which is shown to be faster than the one used in RAPID.

References

[Bell et al. 97] G. Bell, R. Carey, and C. Marrin. VRML97: The Virtual Reality Modeling Language. http://www.vrml.org/Specifications/VRML97, 1997.

[Gottschalk 96a] S. Gottschalk. RAPID: Robust and Accurate Polygon Interference Detection System. http://www.cs.unc.edu/~geom/OBB/OBBT.html, 1996. Software library.

[Gottschalk 96b] S. Gottschalk. "Separating Axis Theorem." Technical Report TR96-024, Department of Computer Science, University of North Carolina at Chapel Hill, 1996.

[Gottschalk et al. 96] S. Gottschalk, M. C. Lin, and D. Manocha. "OBBTree: A Hierarchical Structure for Rapid Interference Detection." In *Computer Graphics (Proc. SIGGRAPH 96)*, edited by Holly Rushmeier, pp. 171–180. New York: ACM Press, 1996.

[Hubbard 96] P. M. Hubbard. "Approximating Polyhedra with Spheres for Time-Critical Collision Detection." *ACM Transactions on Graphics*, 15(3):179–210 (July 1996).

[Klosowksi et al. 98] J. T. Klosowski, M. Held, J. S. B. Mitchell, H. Sowizral, and K. Zikan. "Efficient Collision Detection Using Bounding Volume Hierarchies of k-DOPs." *IEEE Transactions on Visualization and Computer Graphics*, 4(1):21–36 (January–March 1998).

[Möller 97] T. Möller. "A Fast Triangle-Triangle Intersection Test." *journal of graphics tools*, 2(2):25–30 (1997).

[Palmer, Grimsdale 95] I. J. Palmer and R. L. Grimsdale. "Collision Detection for Animation Using Sphere-trees." *Computer Graphics Forum*, 14(2):105–116 (1995).

[Shoemake 92] K. Shoemake. "Uniform Random Rotations." In *Graphics Gems III*, edited by D. Kirk, pages 124–132. Boston: Academic Press, 1992.

[Smith et al. 95] A. Smith, Y. Kitamura, H. Takemura, and F. Kishino. "A Simple and Efficient Method for Accurate Collision Detection Among Deformable Polyhedral Objects in Arbitrary Motion." In *Proc. IEEE Virtual Reality Annual International Symposium*, pp. 136–145, (March 1995).

[Weghorst et al. 94] H. Weghorst, G. Hooper, and D. P. Greenberg. "Improved Computational Methods for Ray Tracing." *ACM Transactions on Graphics*, 3(1):52–69 (January 1994).

[Zachmann 98] G. Zachmann. "Rapid Collision Detection by Dynamically Aligned DOP-trees." In *Proc. IEEE Virtual Reality Annual International Symposium*, pp. 90–97 (March 1998).

Web Information:

Information on obtaining the complete C++ source code and documentation for SOLID 2.0 is available online at:
http://www.acm.org/jgt/papers/vandenBergen97

Gino van den Bergen, Department of Mathematics and Computing Science, Eindhoven University of Technology, P.O. Box 513, 5600 MB Eindhoven, The Netherlands (gino@win.tue.nl)

Received June 23, 1998; accepted in revised form November 5, 1998

New Since Original Publication

The validity of the claims made in this paper strongly depends on the underlying computer architecture that executes these algorithms. Since 1998, the year the first edition of this paper was published, some considerable advances in computer processing power have been made. A few years ago, I decided to repeat the performance analysis on computer hardware of that time, and found an overall performance increase that nicely follows Moore's law. The performance of both the OBB tree and AABB tree has increased by an order of magnitude over the five-year period. However, more interesting is the fact that the performance ratios between the OBB tree and the AABB tree have not changed over the years. The results of the latest experiment can be found in [van den Bergen 03]. Despite the fact that CPUs have gained relatively stronger floating-point processing powers and rely more on processor cache, the claims made in this paper are still valid today.

With the advent of hardware graphics accelerators, the complexity of geometric data has outgrown the memory capacity of computers. On platforms that have a very limited amount of memory, such as game consoles, the relatively large memory usage of bounding-volume hierarchies can be a problem. Fortunately, the memory footprint of bounding-box trees can be further reduced by applying aggressive compression techniques. Notably, the AABB tree's footprint can be reduced from 64 bytes to a mere 11 bytes per primitive [Gomez 01]. Compressed bounding-box trees require additional processing for converting the compressed data to a bounding box representation that can be used in intersection tests, so apparently there is a trade-off of performance against memory size. However, on platforms that are bound by memory-access bandwidth rather than processing power, using compressed AABB trees may actually result in a performance increase [Terdiman 01].

Additional References

[Gomez 01] M. Gomez. "Compressed Axis-Aligned Bounding Box Trees." In *Game Programming Gems 2*, edited by M. DeLoura, pp. 388–393. Hingham, MA: Charles River Media, 2001.

[Terdiman 01] P. Terdiman. "Memory-Optimized Bounding-Volume Hierarchies." Available from World Wide Web (http://codercorner.com/CodeArticles.htm), March 2001.

[van den Bergen 03] G. van den Bergen. *Collision Detection in Interactive 3D Environments*. San Francisco, CA: Morgan Kaufmann, 2003.

Updated Web Information:

DTECTA has released the complete C++ source code and documentation for SOLID 3.5 under the GNU General Public License and the Trolltech Q Public License. The latest distribution of SOLID 3.5 can be downloaded from http://www.dtecta.com

Current Contact Information:

Gino van den Bergen, DTECTA, Sterkenstraatje 4, 5708 ZC Helmond, The Netherlands (gino@dtecta.com)

Vol. 8, No. 1: 25–32

Fast and Robust Triangle-Triangle Overlap Test Using Orientation Predicates

Philippe Guigue and Olivier Devillers

Abstract. This paper presents an algorithm for determining whether two triangles in three dimensions intersect. The general scheme is identical to the one proposed by Möller [Möller 97]. The main difference is that our algorithm relies exclusively on the sign of 4×4 determinants and does not need any intermediate explicit constructions which are the source of numerical errors. Besides the fact that the resulting code is more reliable than existing methods, it is also more efficient. The source code is available online.

1. Introduction

Robustness issues in geometric algorithms have been widely studied in recent years [Yap 97]. It appears that a good solution consists in a design that strictly separates the geometric tests (also called *predicates*) from the combinatorial part of the algorithm. In that way, the robustness issues due to numerical inaccuracy are isolated in the predicates and can be solved using relevant arithmetic tools.

From this point of view, a predicate is a piece of code which answers a particular basic geometric question. This code usually has two independent aspects. The first aspect is an algebraic formulation of the problem, most often, an algebraic expression whose sign gives the answer to the predicate. The second aspect is the arithmetic used to evaluate the value of that expression: It can be rounded arithmetic which is fast, but inexact, or a more elaborate arithmetic which allows us to certify the exactness of the result.

Several quantities can be used to evaluate the quality of a predicate: the degree of the algebraic expression, the number of operations, or the running time. The arithmetic degree of the predicate [Liotta et. al. 98] is the highest degree of the polynomials that appear in its algebraic description. A low-degree predicate has two advantages: Used with an approximate arithmetic, it has a better precision and used with an exact arithmetic, fewer bits are required to represent numbers.

In this paper, we propose a new formulation of the predicate of the intersection of two triangles in three dimensions. This formulation can be decomposed in a small number of three-dimensional orientation tests and has only degree three. This is a big improvement compared to the most popular implementations of this predicate due to Möller [Möller 97] and Held [Held 97] which both have degree at least seven. Using IEEE floating-point computation, this new formulation improves the running time by around 10-20% while improving the stability of the algorithm. The source code is available online at the web site listed at the end of this paper.

The structure of this paper is as follows: In Section 2, we introduce notations and shortly discuss the algorithms proposed by Möller and Held. Section 3 presents the new formulation of the overlap test. Finally, Section 4 compares the performance of the three algorithms.

2.   Notations and Related Work

2.1.   Definition and Notations

Definition 1. *Given four three-dimensional points* $a = (a_x, a_y, a_z)$, $b = (b_x, b_y, b_z)$, $c = (c_x, c_y, c_z)$, *and* $d = (d_x, d_y, d_z)$, *we define the determinant*

$$[a, b, c, d] := - \begin{vmatrix} a_x & b_x & c_x & d_x \\ a_y & b_y & c_y & d_y \\ a_z & b_z & c_z & d_z \\ 1 & 1 & 1 & 1 \end{vmatrix} = (d - a) \cdot ((b - a) \times (c - a)). \quad (1)$$

The sign of $[a, b, c, d]$ has two geometric interpretations, each corresponding to a right-hand rule. It tells whether vertex d is above, below, or on a plane through a, b, and c, where *above* is the direction of a right-handed screw at a that turns from b toward c. Equivalently, it tells whether a screw directed along the ray $\overrightarrow{ab}$ turns in the direction of $\overrightarrow{cd}$ (see Figure 1(a) and 1(b)). In either interpretation, the result is zero iff the four points are coplanar.

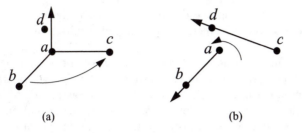

(a)   (b)

Figure 1. $[a, b, c, d]$ is positive according to a right-hand rule for planes or for lines.

In the sequel, T_1 and T_2 denote triangles with vertices p_1, q_1, and r_1, and p_2, q_2, r_2, respectively. π_1 and π_2 denote their respective supporting planes, and L the line of intersection of the two planes (see Figure 2). The overlap test described in this paper returns a boolean value which is true if the closed triangles (i.e., the triangles including their boundary) intersect. The algorithm does not handle degenerate triangles (i.e., line segments and points).

2.2. Related Work

Several solutions exist to test the intersection between three-dimensional triangles, although the most important ones, due to their efficiency, are the algorithms proposed by Möller [Möller 97] and Held [Held 97]. However, both are based on a constructive multistep process (that is, both methods have to explicitly construct some intersection points) which has two disadvantages. First, degenerate cases at each step of the process need to be handled

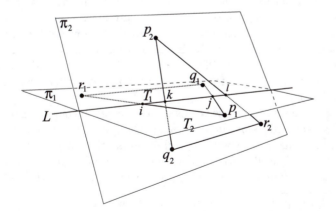

Figure 2. Triangles and the planes in which they lie. After Möller [Möller 97].

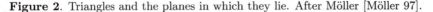

specifically. Second, using the derived values from the initial constructions increases the required arithmetic precision and results in an error-prone implementation.

The triangle intersection test presented in this paper is similar to Möller's algorithm [Möller 97], but relies exclusively on the sign of orientation predicates. In consequence, its implementation is both faster (it uses fewer arithmetic operations) and more robust (it keeps the accumulation of roundoff errors small by using fewer consecutive multiplication operations) than constructive process approaches. Section 3 details the algorithm.

3. Three-Dimensional Triangle-Triangle Overlap Test

Each triangle is a subset of the plane it lies in, so for two triangles to intersect, they must overlap along the line of intersection of their planes. Hence, a necessary condition for intersection is that each triangle must intersect the plane of the other. Based on this remark, some configurations of nonintersecting triangle pairs can be quickly detected by checking if one triangle lies entirely in one of the open halfspaces induced by the supporting plane of the other.

The algorithm begins similarly as Möller's code [Möller 97] by checking the mutual intersection of each triangle with the plane of the other.[1] To do so, the algorithm first tests the intersection of triangle T_1 with the plane π_2 by comparing the signs of $[p_2, q_2, r_2, p_1]$, $[p_2, q_2, r_2, q_1]$, and $[p_2, q_2, r_2, r_1]$. Three distinct situations are possible: If all three determinants evaluate to the same sign and no determinant equals zero, then all three vertices of T_1 lie on the same open halfspace induced by π_2. In this case, no further computation is needed since there can be no intersection. If all three determinants equal zero, then both triangles are co-planar. This special case is solved by a two-dimensional triangle-triangle overlap test after projecting the three-dimensional input vertices onto the axis-aligned plane where the areas of the triangles are maximized.[2] Finally, if the determinants have different signs, then the vertices of T_1 lie on different sides of the plane π_2 and T_1 surely intersects π_2. In this last case, the algorithm then checks whether T_2 intersects π_1 in a similar manner.

Assume now that no edge nor vertex of one triangle belongs to L, otherwise, suppose that such edges or vertices are slightly perturbed to one side of the other triangle's plane in a way that the intersection of each triangle with L is preserved. The algorithm then applies a circular permutation to the vertices of each triangle such that p_1 (respectively, p_2) is the only vertex

---

[1] The ERIT code [Held 97] only checks whether T_2 intersects π_1.

[2] The two-dimensional routine provided in the source code is an implementation of the algorithm in [Devillers and Guigue 02] that performs ten 2×2 determinant sign evaluations in the worst case.

of its triangle that lies on its side. An additional transposition operation (i.e., a swap operation) is performed at the same time on vertices q_2 and r_2 (respectively, q_1 and r_1) so that vertex p_1 (respectively, p_2) sees $p_2 q_2 r_2$ (respectively, $p_1 q_1 r_1$) in counterclockwise order (see Figure 2).

Due to our previous rejections and permutations, each of the incident edges of the vertices p_1 and p_2 is now guaranteed to intersect L at a unique point. Let i, j, k, and l, be scalar values that represent the intersection points of L with edges $p_1 r_1$, $p_1 q_1$, $p_2 q_2$, and $p_2 r_2$, respectively. These values form closed scalar intervals I_1 and I_2 on L that correspond to the intersection between the two input triangles and L. Furthermore, at this step, there is enough information to know a consistent order of the bounds of each interval. Precisely, $I_1 = [i, j]$ and $I_2 = [k, l]$ if L is oriented in the direction of $N = N_1 \times N_2$ where $N_1 = (q_1 - p_1) \times (r_1 - p_1)$ and $N_2 = (q_2 - p_2) \times (r_2 - p_2)$. Thus, it is only necessary to check a min/max condition to determine whether or not the two intervals overlap. By min/max condition, we mean that the minimum of either interval must be smaller than the maximum of the other, namely, $k \leq j$ and $i \leq l$. Since the edges $p_1 r_1$, $p_1 q_1$, $p_2 q_2$, and $p_2 r_2$ intersect L, this condition reduces to check the predicate

$$[p_1, q_1, p_2, q_2] \leq 0 \ \wedge \ [p_1, r_1, r_2, p_2] \leq 0 \ . \tag{2}$$

This can be easily seen by looking at Figure 2 and recalling the second interpretation of the three-dimensional orientation test. Indeed, this conjunction amounts to test whether a screw directed along the ray $\overrightarrow{p_1 q_1}$ (respectively, $\overrightarrow{p_1 r_1}$) turns in the direction of $\overrightarrow{q_2 p_2}$ (respectively, $\overrightarrow{p_2 r_2}$). Hence, the explicit construction of intermediate objects that is needed in [Möller 97] is avoided by recasting the ordering of the projections of the triangles' vertices onto L as the evaluation of the sign of at most two determinants of degree three in the input variables.

A description of the actual line of intersection of the triangles can be computed either by replacing the two last orientation tests by three other tests or by performing two more orientation tests at the end. These tests determine a complete ordering of i, j, k, and l on L which allows us to deduce the two scalars that form the interval $I_1 \cap I_2$. The two endpoints of the line segment of intersection can then be constructed in terms of edge-triangle intersections defined by each interval bound.

4.   Comparative Analysis

The algorithm has been implemented using C language and operates on the double data type. The code does not handle degenerate triangles (i.e., line segments and points). However, if one wishes to handle those cases, they must be detected (all tests involving a degenerate triangle equal zero) and handled as special cases. Note that if triangles are assumed to be nondegenerate, coplanarity of the triangles can be detected as soon as the first three orientation tests are computed (all three signs equal zero).

In the following, we give the arithmetic degrees and the running times of the different algorithms. We first compare our code in terms of arithmetic requirements to the two fastest existing methods due to Möller and Held.[3]

4.2.   Arithmetic degrees

In our algorithm, all branching decisions are carried out by evaluating the sign of orientation predicates which are degree three polynomials. Therefore, the algorithm has degree three. To compare this against the Möller approach, we could parameterize the line of intersection of the two supporting planes as $L(t) = O + tD$, where D is the cross product of the normals to the plane and O is some point on it, and solve for the values of t (i.e., scalar intervals on L) that correspond to the intersection between the triangles and L. The algorithm then has to compare these t values. There actually exist two different implementations of the algorithm in [Möller 02]. tri_tri_intersect uses a simplified projection of the vertices onto L and has to compare rational polynomials with numerator of degree four and denominator of degree three. However, this version contains divisions and shall be seen as a real polynomial implementation of degree seven. NoDivTriTriIsect replaces divisions with multiplications and is a little faster, but has to compare degree thirteen polynomials. Held's algorithm computes the line segment of intersection of T_2 with the plane π_1 by evaluating rational polynomial with numerator of degree four and denominator of degree three. It then solves a two-dimensional segment-triangle overlap test that involves degree two polynomials in the input variables. Hence, all three algorithms have degree at least seven. Consequently, our algorithm allows us to limit the arithmetic requirements of a robust implementation. When operations are implemented in IEEE754 floating-point arithmetic (as originally done in Möller and Held codes), then fewer bits are lost due to rounding and all branching decisions are proved to be correct with high probability if the triangles vertices are independently evenly distributed in a cube [Devillers and Preparata 98]. When operations are implemented exactly using a multiprecision package or fixed point, then our method requires less arithmetic or allows a larger domain for input.

4.3.   Optimizations

Note that the computation of the first six determinants can be optimized by computing the invariant parts consisting of 2×2 subdeterminants only once.

---

[3]Möller's implementation was taken from the web site [Möller 02]. The code for ERIT Version 1.1. was obtained from its author.

Code	(1)		(2)		(3)		(4)	
	μ sec	%	μ sec	%	μ sec	%	μ sec	%
Our tri_tri_overlap_test()	0.34	0	0.18	0	0.33	10	0.32	41
Möller's NoDivTriTriIsect()	0.46	0	0.27	0	0.46	33	0.46	53
Möller's tri_tri_intersect()	0.53	0	0.28	0	0.53	31	0.54	54
Held's TriTri3D()	0.48	0	0.30	0	0.48	43	0.47	29
Our tri_tri_intersection_test()	0.55	0	0.19	0	0.53	9	0.43	40
Möller's tri_tri_intersect_with_isectline()	0.62	0	0.30	0	0.61	31	0.60	54

Table 1. CPU times (in microseconds) and failure percentages of the algorithms using IEEE double arithmetic on (1) random intersecting, (2) random nonintersecting, (3) degenerate intersecting, and (4) degenerate nonintersecting triangle pairs.

For $[a, b, c, X]$, expansion by 2×2 subdeterminants gives the normal vector of the plane passing through a, b, and c, and reduces the determinants to a dot product of 3-tuples.

4.4. Performance

We tested the performance and the robustness of the three methods for different kinds of data in general and degenerate position (see Table 1). For *random* input, we use points whose coordinate values are chosen pseudo-randomly using the `drand48()` system call. The *degenerate* input are triangle pairs with grazing contacts obtained by generating two coplanar edges that intersect and two points on each side of the plane and converting exact coordinates to double precision. We compared our code `tri_tri_overlap_test()` to the two implementations of Möller's algorithm and to Held's code `TriTri3D()`. The table also gives timings for Möller's `tri_tri_intersect_with_isectline()` and our `tri_tri_intersection_test()` which are versions that compute the segment of intersection. The entries of the table were computed by normalizing the timings recorded from running several millions of intersection tests. Our overlap test was found to be between 10 and 20 percent faster than the fastest of the other methods in all cases when using IEEE double-precision, floating-point computation. If exactness is necessary, our algorithm can benefit from speed-up techniques [Shewchuk 97, Brönnimann et. al. 99] to minimize the overhead of exact arithmetic. This is not possible in approaches that use intermediate constructions.

References

[Brönnimann et. al. 99] Hervé Brönnimann, Ioannis Emiris, Victor Pan, and Sylvain Pion. "Sign Determination in Residue Number Systems." *Theoret. Comput. Sci.* 210:1 (1999), 173–197.

[Devillers and Preparata 98] O. Devillers and F. P. Preparata. "A Probabilistic Analysis of the Power of Arithmetic Filters." *Discrete and Computational Geometry* 20 (1998), 523–547.

[Devillers and Guigue 02] Olivier Devillers and Philippe Guigue. "Faster Triangle-Triangle Intersection Tests." Rapport de recherche 4488, INRIA, 2002.

[Held 97] Martin Held. "ERIT – A Collection of Efficient and Reliable Intersection Tests." *journal of graphics tools* 2:4 (1997), 25–44.

[Liotta et. al. 98] Giuseppe Liotta, Franco P. Preparata, and Roberto Tamassia. "Robust Proximity Queries: An Illustration of Degree-Driven Algorithm Design." *SIAM J. Comput.* 28:3 (1998), 864–889.

[Möller 97] Tomas Möller. "A Fast Triangle-Triangle Intersection Test." *journal of graphics tools* 2:2 (1997), 25–30.

[Möller 02] Tomas Möller. "A Fast Triangle-Triangle Intersection Test." Available from World Wide Web (http://www.acm.org/jgt/papers/Moller97/), 2002.

[Shewchuk 97] Jonathan Richard Shewchuk. "Adaptive Precision Floating-Point Arithmetic and Fast Robust Geometric Predicates." *Discrete Comput. Geom.* 18:3 (1997), 305–363.

[Yap 97] C. Yap. "Towards Exact Geometric Computation." *Comput. Geom. Theory Appl.* 7:1 (1997), 3–23.

Web Information:

A sample C implementation of our algorithm is available at
http://www.acm.org/jgt/papers/GuigueDevillers03/.

Philippe Guigue, INRIA Sophia Antipolis, BP 93, 2004 Route des Lucioles, 06902 Sophia Antipolis Cedex, France (philippe.guigue@sophia.inria.fr)

Olivier Devillers, INRIA Sophia Antipolis, BP 93, 2004 Route des Lucioles, 06902 Sophia Antipolis Cedex, France (olivier.devillers@sophia.inria.fr)

Received August 21, 2002; accepted in revised form January 11, 2003.

New Since Original Publication

We briefly describe in the following how the value of determinants introduced so far can be used to obtain an efficient solution for the ray-triangle intersection problem. The structure of the obtained intersection test is identical to the algorithm proposed by Möller and Trumbore [Möller, Trumbore 97][4]. It performs the same rejection tests based on the barycentric coordinates of the intersection point between the ray and the triangle's supporting plane. The algorithms differ only in the way the tested values are computed. Our new formulation allows optimizations in the case that the triangle normal is known.

More precisely, let $\mathbf{abc}$ be a triangle with normal $\mathbf{n}$, and let $r(t) = \mathbf{o} + t\mathbf{d}$ be a ray defined by its origin point $\mathbf{o}$ and a direction vector $\mathbf{d}$. One can show that the determinants $[\mathbf{o}, \mathbf{d}, \mathbf{a}, \mathbf{c}]$ and $-[\mathbf{o}, \mathbf{d}, \mathbf{a}, \mathbf{b}]$ are proportional to the barycentric coordinates (u, v) of the ray-plane intersection point, $\mathbf{r}(t) = (1 - u - v)\mathbf{a} + u\mathbf{b} + v\mathbf{c}$ (cf. [Jones 00]). Moreover, $[\mathbf{a}, \mathbf{b}, \mathbf{c}, \mathbf{o}]$ is proportional to its ray parameter t, and the proportionality constant for all three values can be shown to be $-[\mathbf{a}, \mathbf{b}, \mathbf{c}, \mathbf{d}]$.

By computing subdeterminants of the two first determinants only once as described in Section 4.3, and since $[\mathbf{a}, \mathbf{b}, \mathbf{c}, \mathbf{o}] = \mathbf{p}.\mathbf{n}$ and $[\mathbf{a}, \mathbf{b}, \mathbf{c}, \mathbf{d}] = \mathbf{d}.\mathbf{n}$, our intersection algorithm reduces to one cross product and four dot products when the normal is given. This improves by a cross product the algorithm described in [Möller, Trumbore 97] which does not presume that normals are precomputed but which cannot, to our knowledge, take advantage of them when they are supplied.

A more detailed description of the algorithm as well as the source code are available online at the paper's website.

Additional References

[Jones 00] Ray Jones. "Intersecting a Ray and a Triangle with Plücker Coordinates." *Ray Tracing News* 13:1 (July 2000). Available from World Wide Web (http://www.acm.org/tog/resources/RTNews/html/rtnv13n1.html)

[Möller, Trumbore 97] Tomas Möller and Ben Trumbore. "Fast, Minimum Storage Ray-Triangle Intersection." *journal of graphics tools* 2:1 (1997), 21–28.

Current Contact Information:

Philippe Guigue, Tools and Toys, GmbH, Brunnenstrasse 196, 10119 Berlin (philippe.guigue@vrtnt.com)

Olivier Devillers, INRIA Sophia Antipolis, BP 93, 2004 Route des Lucioles, 06902 Sophia Antipolis Cedex, France (olivier.devillers@sophia.inria.fr)

---

[4]See also pages 181–189 in this book.

Vol. 6, No. 2: 43–52

A Simple Fluid Solver Based on the FFT

Jos Stam

Alias | wavefront

Abstract. This paper presents a very simple implementation of a fluid solver. The implementation is consistent with the equations of fluid flow and produces velocity fields that contain incompressible rotational structures and dynamically react to user-supplied forces. Specialized for a fluid which wraps around in space, it allows us to take advantage of the Fourier transform, which greatly simplifies many aspects of the solver. Indeed, given a Fast Fourier Transform, our solver can be implemented in roughly one page of readable C code. The solver is a good starting point for anyone interested in coding a basic fluid solver. The fluid solver presented is useful also as a basic motion primitive that can be used for many different applications in computer graphics.

1. Introduction

Simulating fluids is one of the most challenging problems in engineering. In fact, over the last 50 years much effort has been devoted to writing large fluid simulators that usually run on expensive supercomputers. Simulating fluids is also important to computer graphics, where there is a need to add visually convincing flows to virtual environments. However, while large production houses may be able to afford expensive hardware and large fluid solvers, most computer graphics practitioners require a simple solution that runs on a desktop PC. In this paper we introduce a fluid solver that can be written in roughly one page of C code. The algorithm is based on our stable fluid solver [Stam 99]. We specialize the algorithm for the case of a periodic domain, where

155

the fluid's boundaries wrap around. In this case, we can use the machinery of Fourier transforms to efficiently solve the equations of fluid flow. Although periodic flows do not exist in nature, they can be useful in many applications. For example, the resulting fluid flow can be used as a "motion texture map." Similar texture maps based on the FFT have been used in computer graphics to model fractal terrains [Voss 88], ocean waves [Mastin et al. 87, Tessendorf 00], turbulent wind fields [Shinya, Fournier 92, Stam, Fiume 93] and the stochastic motion of flexible structures [Stam 97], for example. Also, our solver is a good starting point for someone new to fluid dynamics, since the computational fluid dynamics literature and the complexity of most existing solvers can be quite intimidating to a novice. We warn the reader, however, that our method sacrifices accuracy for speed and stability: our simulations are damped more rapidly than an accurate flow. On the other hand, our simulations do converge to the solutions of the equations of fluid flow in the limit when both the grid spacing and the time step go to zero.

2.   Basic Structure of the Solver

The class of fluids that we will simulate in this paper is entirely described by a velocity field that evolves under the action of external forces. The velocity field is defined on a regular grid of N^d voxels, where d is the dimension of the domain. To each voxel's center we assign a velocity. Typically, we consider only two-dimensional and three-dimensional flows. However, our algorithm also works for flows living in an arbitrary d-dimensional space. For the sake of clarity, we describe our algorithm in a two-dimensional space where the underlying concepts are easier to visualize. We assume that our grid has the topology of a torus: the fluid is continuous across opposing boundaries. Figure 1 illustrates this situation. Because the velocity is periodic, it can be used to tile the entire plane seamlessly. Although such fluids do not occur in nature, they can be useful in computer graphics as a motion texture map defined everywhere in space.

Our solver updates the grid's velocities over a given time step, dt. Consequently the simulation is a set of snapshots of the fluid's velocity over time. The simulation is driven by a user-supplied grid of external forces and a viscosity, visc. The evolution of the fluid is entirely determined by these parameters. Subsequently, at each time step the fluid's grid is updated in four steps:

- Add force field,

- Advect velocity,

- Diffuse velocity,

- Force velocity to conserve mass.

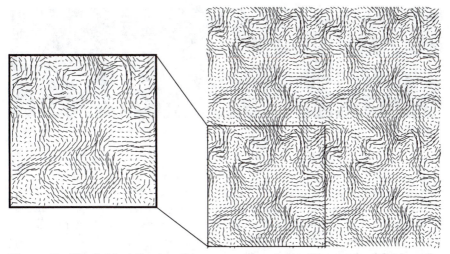

Figure 1. The fluids defined in this paper are periodic. The patch of fluid on the left can be used to seamlessly tile the entire plane.

The basic idea behind our solver is to perform the first two steps in the spatial domain and the last two steps in the Fourier domain. Before describing each step in more detail, we will say more about the Fourier transform applied to velocity fields.

3. A Fluid in the Fourier Domain

The main reason to introduce the Fourier transform is that many operations become very easy to perform in the Fourier domain. This fact is well known in image and signal processing, where convolution becomes a simple multiplication in the Fourier domain. Before introducing Fourier transforms of vector fields, we outline the basic ideas for single-valued scalar functions. For a more thorough treatment of the Fourier transform, the reader is referred to any of the standard textbooks, e.g., [Bracewell 78]. Informally, a two-dimensional function can be written as a weighted sum of planar waves with different wavenumbers. In Figure 2 we depict four planar waves with their corresponding wavenumbers, $\mathbf{k} = (k_x, k_y)$ (shown in white).

The direction of the wavenumber is perpendicular to the crests of the wave, while the length of the wavenumber, $k = \sqrt{k_x^2 + k_y^2}$, is inversely proportional to the spacing between the waves. Larger wavenumbers therefore correspond to higher spatial frequencies. Any periodic two-dimensional function can be represented as a weighted sum of these simple planar waves. The Fourier

Vol. 6, No. 2: 43–52

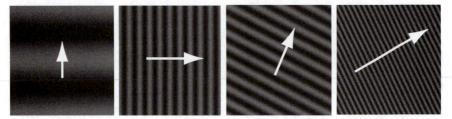

Figure 2. Four planar waves and their corresponding wavenumbers (white arrows).

transform of the signal is defined as a function that assigns the corresponding weight to each wavenumber **k**. In Figure 3 we depict two pairs of functions with their corresponding Fourier transform. Notice that the smoother function on the left has a Fourier transform with weights that decay more rapidly for higher spatial frequencies than the one on the right.

To take the Fourier transform of a two-dimensional vector field, we independently take the Fourier transformations of the u components of the vectors and the v components of the vectors, each considered as a two-dimensional scalar field. Figure 4 shows a two-dimensional velocity field in the spatial domain (left) and the corresponding Fourier transform (right). The wavenumbers take values in the range $[-N/2, N/2] \times [-N/2, N/2]$, so that the origin is at the center. Since the velocity field is relatively smooth, we observe that the vectors are smaller for large wavenumbers, which is a common feature of most velocity fields. This observation directly suggests a method to account for the effects of viscosity which tend to smooth out the velocity field. Simply multiply the Fourier transform by a filter which decays along with the spatial frequency, the viscosity and the time step (see Section 4.3).

As mentioned in the previous section not all operations are best performed in the Fourier domain. Therefore, we need a mechanism to switch between the spatial and the Fourier domain. Fortunately this can be done very efficiently using a Fast Fourier Transform (FFT). We will not describe an implementation of the FFT in this paper because very efficient public domain "black box"

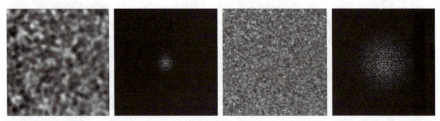

Figure 3. Two examples of a two-dimensional function and its corresponding FFT. The pair on the left is smoother than the one on the right.

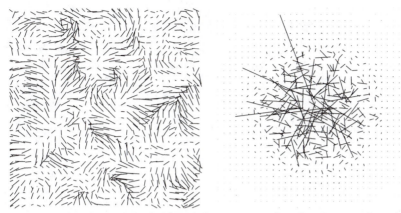

Figure 4. A velocity field (left) and the corresponding Fourier transform (right).

solvers are readily available. The one used in this paper is MIT's FFTW, the "Fastest Fourier Transform in the West."[1]

4. The Solver

We now describe in more detail the four steps of our solver. In Figure 5 we provide the corresponding implementation in C. Numbers between brackets in this section refer to the corresponding lines in the C code. The input to our solver is the velocity of the previous time step (u,v), the time step dt, a force field defined on a grid (u0,v0) and the viscosity of the fluid visc [17-18]. Because of the slight overhead in memory associated with the FFT, the arrays u0 and v0 should be of size n*(n+2). Prior to calling the solver, the routine init_FFT should be called at least once.

4.1. Adding Forces

This is the easiest step of our solver: we simply add the force grids multiplied by the time step to the velocity grid [23-26]. Consequently, in regions where the force field is large the fluid will move more rapidly.

```
01 #include <srfftw.h>
02
03 static rfftwnd_plan plan_rc, plan_cr;
04
05 void init_FFT ( int n )
06 {
07     plan_rc = rfftw2d_create_plan ( n, n, FFTW_REAL_TO_COMPLEX, FFTW_IN_PLACE );
```

Figure 5. Our fluid solver coded in C.

---

[1] Available at http://www.fftw.org

```
08    plan_cr = rfftw2d_create_plan ( n, n, FFTW_COMPLEX_TO_REAL, FFTW_IN_PLACE );
09  }
10
11  #define FFT(s,u)\
12  if (s==1) rfftwnd_one_real_to_complex ( plan_rc, (fftw_real *)u, (fftw_complex *)u );\
13  else rfftwnd_one_complex_to_real ( plan_cr, (fftw_complex *)u, (fftw_real *)u )
14
15  #define floor(x) ((x)>=0.0?((int)(x)):(-((int)(1-(x)))))
16
17  void stable_solve ( int n, float * u, float * v, float * u0, float * v0,
18  float visc, float dt )
19  {
20    float x, y, x0, y0, f, r, U[2], V[2], s, t;
21    int i, j, i0, j0, i1, j1;
22
23    for ( i=0 ; i<n*n ; i++ ) {
24      u[i] += dt*u0[i]; u0[i] = u[i];
25      v[i] += dt*v0[i]; v0[i] = v[i];
26    }
27
28    for ( x=0.5/n,i=0 ; i<n ; i++,x+=1.0/n ) {
29      for ( y=0.5/n,j=0 ; j<n ; j++,y+=1.0/n ) {
30        x0 = n*(x-dt*u0[i+n*j])-0.5; y0 = n*(y-dt*v0[i+n*j])-0.5;
31        i0 = floor(x0); s = x0-i0; i0 = (n+(i0%n))%n; i1 = (i0+1)%n;
32        j0 = floor(y0); t = y0-j0; j0 = (n+(j0%n))%n; j1 = (j0+1)%n;
33        u[i+n*j] = (1-s)*((1-t)*u0[i0+n*j0]+t*u0[i0+n*j1])+
34                      s *((1-t)*u0[i1+n*j0]+t*u0[i1+n*j1]);
35        v[i+n*j] = (1-s)*((1-t)*v0[i0+n*j0]+t*v0[i0+n*j1])+
36                      s *((1-t)*v0[i1+n*j0]+t*v0[i1+n*j1]);
37      }
38    }
39
40    for ( i=0 ; i<n ; i++ )
41      for ( j=0 ; j<n ; j++ )
42        { u0[i+(n+2)*j] = u[i+n*j]; v0[i+(n+2)*j] = v[i+n*j]; }
43
44    FFT(1,u0); FFT(1,v0);
45
46    for ( i=0 ; i<=n ; i+=2 ) {
47      x = 0.5*i;
48      for ( j=0 ; j<n ; j++ ) {
49        y = j<=n/2 ? j : j-n;
50        r = x*x+y*y;
51        if ( r==0.0 ) continue;
52        f = exp(-r*dt*visc);
53        U[0] = u0[i  +(n+2)*j]; V[0] = v0[i  +(n+2)*j];
54        U[1] = u0[i+1+(n+2)*j]; V[1] = v0[i+1+(n+2)*j];
55        u0[i  +(n+2)*j] = f*( (1-x*x/r)*U[0]    -x*y/r *V[0] );
56        u0[i+1+(n+2)*j] = f*( (1-x*x/r)*U[1]    -x*y/r *V[1] );
57        v0[i+  (n+2)*j] = f*(   -y*x/r *U[0] + (1-y*y/r)*V[0] );
58        v0[i+1+(n+2)*j] = f*(   -y*x/r *U[1] + (1-y*y/r)*V[1] );
59      }
60    }
61
62    FFT(-1,u0); FFT(-1,v0);
63
64    f = 1.0/(n*n);
65    for ( i=0 ; i<n ; i++ )
66      for ( j=0 ; j<n ; j++ )
67        { u[i+n*j] = f*u0[i+(n+2)*j]; v[i+n*j] = f*v0[i+(n+2)*j]; }
68  }
```

Figure 5 (continued). Our fluid solver coded in C.

4.2. Self-Advection

This step accounts for the non-linearities present in a fluid flow which makes them unpredictable and chaotic. Fluids such as air can advect (transport) many substances such as dust particles. In the latter case the particle's velocity is equal to that of the fluid at the particle's position. On the other hand, a fluid is also made up of matter. The particles that constitute this matter are also advected by the fluid's velocity. Over a time step these particles will move and transport the fluid's velocity to another location. This effect is called "self-advection": an auto-feedback mechanism within the fluid. To solve for this effect we use a very elegant technique known as a "semi-Lagrangian" scheme in the computational fluid dynamics literature [28-38].

Here is how it works: For each voxel of the velocity grid, we trace its midpoint backwards through the velocity field over a time step, dt. This point will end up somewhere else in the grid. We then linearly interpolate the velocity at that point from the neighboring voxels and transfer this velocity back to the departure voxel. This requires two grids, one for the interpolation (u0,v0) and one to store the new interpolated values (u,v). To perform the particle trace, we use a simple linear approximation of the exact path. The interpolation is very easy to implement because our grids are periodic. Points that leave the grid simply re-enter the grid from the opposite side.

4.3. Viscosity

We alluded to this step in Section 3. We first transform our velocity field into the Fourier domain [44]. We then filter out the higher spatial frequencies by multiplying the Fourier transform by a filter whose decay depends on the magnitude of wavenumber, the viscosity and the time step [46-60]. A filter with all these properties which is consistent with the equations of fluid flow is given by exp(-k^2*visc*dt). Since we are only interested in visual accuracy, other, possibly cheaper, filters could be used.

4.4. Conservation of Mass

Most fluids encountered in practice conserve mass. Looking at a small piece of fluid, we expect the flow coming in to be equal to the flow going out. However, after the previous three steps this is rarely the case. We will correct this in a final step. As for viscosity, it turns out that this step is best performed in the Fourier domain. To solve for this step, we use a mathematical result known as the "Helmholtz decomposition" of a velocity field: every velocity field is the sum of a mass-conserving field and a gradient field. A gradient

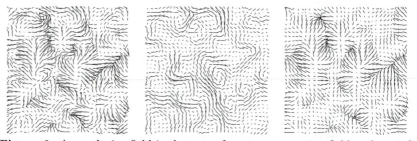

Figure 6. Any velocity field is the sum of a mass conserving field and a gradient field.

field is one that comes from a scalar field and indicates at each point in which direction the scalar field increases most. In two dimensions the scalar field can be visualized as a height field. In this case, the gradient field simply points to the steepest upward slope. Figure 6 depicts the decomposition for an arbitrary field shown on the left. The middle field is the mass-conserving field while the right most one is the gradient field. Notice the amount of swirliness in the mass-conserving field. Visually, this is clearly the type of behavior that we are seeking in computer graphics. The gradient field, on the other hand, is the worst case scenario: at almost every point the flow is either completely in-flowing or completely out-flowing.

In Figure 7 we show the corresponding fields in the Fourier domain. Notice that these fields have a very simple structure. The velocity of the mass conserving field is always perpendicular to the corresponding wavenumber, while the gradient field is always parallel to its wavenumber. Therefore, in the Fourier domain we force our field to be mass conserving by simply projecting each velocity vector onto the line (or plane in three dimensions) perpendicular to the wavenumber [53-58]. This is a very simple operation. The mathematical reason for the simplicity of the mass-conserving and gradient fields in the Fourier domain follows from the fact that differentiation in the spatial domain corresponds to a multiplication by the wavenumber in the Fourier domain.

Finally after this step we transform the velocity field back into the spatial domain [62] and normalize it [64-67].

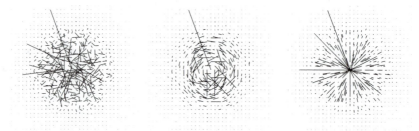

Figure 7. The corresponding decomposition of Figure 6 in the Fourier domain.

4.5. Parameters

Typical values for the parameters of the fluid solver are `dt=1`, `visc=0.001` and force values with a magnitude of roughly 10. These values give nice animations of fluids.

5. Conclusions

We have shown that a simple fluid solver can be coded in roughly one page of C code. Our solver, as shown in [Stam 99] can be advanced at any speed and is unconditionally stable. This is highly desirable in computer graphics applications. Although the code is only given for a two-dimensional fluid, it can easily be extended to handle three-dimensional or even higher dimensional fluids. Applications of fluid dynamics to dimensions higher than three are unknown to the author. Perhaps the velocity (rate of change) of a set of d parameters might be governed by a fluid equation. We leave this as a topic for future research.

References

[Bracewell 78] Ronald N. Bracewell. *The Fourier Transform and its Applications*, 2nd edition. New York: McGraw-Hill, 1978.

[Mastin et al. 87] Gary A. Mastin, Peter A. Watterberg, and John F. Mareda. "Fourier Synthesis of Ocean Scenes." *IEEE Computer Graphics and Applications* 7(3): 16–23 (March 1987).

[Shinya, Fournier 92] Mikio Shinya and Alain Fournier. "Stochastic Motion–Motion Under the Influence of Wind." In *Computer Graphics Forum (Eurographics '92)*, 11(3), pp. 119–128, September 1992.

[Stam 97] Jos Stam. "Stochastic Dynamics: Simulating the Effects of Turbulence on Flexible Structures." *Computer Graphics Forum (Eurographics '97 Proceedings)*, 16(3): 159–164, 1997.

[Stam 99] Jos Stam. "Stable Fluids." In *Proceedings of SIGGRAPH 99, Computer Graphics Proceedings, Annual Conference Series*, edited by Alyn Rockwood, pp. 121–128, Reading, MA: Addison Wesley Longman, 1999.

[Stam, Fiume 93] Jos Stam and Eugene Fiume. "Turbulent Wind Fields for Gaseous Phenomena." In *Proceedings of SIGGRAPH 93, Computer Graphics Proceedings, Annual Conference Series*, edited by James T. Kajiya, pp. 369–376, New York: ACM Press, 1993.

[Tessendorf 00] Jerry Tessendorf. "Simulating Ocean Water." In *SIGGRAPH '2000: Course Notes 25: Simulating Nature: From Theory to Practice*. New York: ACM SIGGRAPH, 2000.

[Voss 88] R. F. Voss. "Fractals in Nature: From Characterization to Simulation." In *The Science of Fractal Images*, pp. 21–70. New York: Springer-Verlag, 1988.

Web Information:

The C code in this paper is available at
http://www.acm.org/jgt/papers/Stam01.

Jos Stam, Alias | wavefront, 1218 Third Avenue, Suite 800, Seattle, WA 98101 (jstam@aw.sgi.com)

Received March 26, 2001; accepted August 2, 2001.

New Since Original Publication

Since the publication of this paper the author has developed an even simpler implementation of the fluid solver which does not rely on a Fast Fourier Transform. This new solver is fully described in [Stam 03] and the source code can be found on the jgt web site.

Additional References

[Stam 03] Jos Stam. "Real-Time Fluid Dynamics for Games." In *Proceedings of the Game Developer Conference, March 2003*. San Francisco: CMP Media, 2003.

Current Contact Information:

Jos Stam, Alias, 210 King Street East, Toronto, Ontario, M5A 1J7, Canada (jstam@alias.com)

Part IV

Ray Tracing

Vol. 3, No. 2: 1–14

Efficiency Issues for Ray Tracing

Brian Smits
University of Utah

Abstract. Ray casting is the bottleneck of many rendering algorithms. Although much work has been done on making ray casting more efficient, most published work is high level. This paper discusses efficiency at a slightly lower level, presenting optimizations for bounding-volume hierarchies that many people use but are rarely described in the literature. A set of guidelines for optimization are presented that avoid some of the common pitfalls. Finally, the effects of the optimizations are shown for a set of models.

1. Introduction

Many realistic rendering systems rely on ray-casting algorithms for some part of their computation. Often, the ray casting takes most of the time in the system, and significant effort is usually spent on making it more efficient. Much work has been done and published on acceleration strategies and efficient algorithms for ray casting, the main ideas of which are summarized in Glassner [Glassner 89]. In addition, many people have developed optimizations for making these algorithms even faster. Much of this work remains unpublished and part of oral history. This paper is an attempt to write down some of these techniques and some higher level guidelines to follow when trying to speed-up ray-casting algorithms. I learned most of the lessons in here the hard way, either by making the mistakes myself, or by tracking them down in other systems. Many of the observations in here were confirmed by others.

This paper will discuss some mid-level optimization issues for bounding volume hierarchies. The ray-casting algorithm uses the hierarchy to determine if the ray intersects an object. An intersection involves computing the distance to the intersection and the intersection point as well as which object was hit. Sometimes it includes computing surface normal and texture coordinates. The information computed during an intersection is sometimes called the hit information. In ray-tracing based renderers, rays from the eye are called primary rays. Reflected and transmitted rays are known as secondary rays. Together, these rays are called intersection rays. Rays from hits to lights to determine shadowing are called shadow rays.

2.   Principles of Optimization

Optimization can be a seductive activity leading to endless tweaks and changes of code. The most important part of optimization is knowing when not to do it. Two common cases are:

- Code or system is not run frequently.

- Code is a small fraction of overall time.

In other words, code should only be optimized if it will make a significant affect on the final system and the final system will be used frequently enough to justify the programmer's time and the chance of breaking something.

It helps to have a set of principles to follow in order to guide the process of optimization. The set I use is:

- Make it work before you make it fast.

- Profile everything you do.

- Complexity is bad.

- Preprocessing is good.

- Compute only what you need.

2.1.   Make it Work Before You Make it Fast

Code should be made correct before it is made fast [Bentley 82]. As stated repeatedly by Knuth [Knuth 92] "Premature optimization is the root of all evil." Obviously, slow, correct code is more useful than fast, broken code. There is an additional reason for the rule, though. If you create a working, unoptimized version first, you can use that as a benchmark to check your optimizations. This is very important. Putting the optimizations in early means you can never be completely sure if they are actually speeding up the code. You don't want to find out months or years later that your code could be sped-up by removing all those clever optimizations.

2.2. Profile Everything You Do

It is important to find out what the bottleneck is before trying to remove it. This is best done by profiling the code before making changes [Bentley 89]. The best profilers give time per-line-of-code as well as per-function. They also tell you how many times different routines are called. Typically what this will tell you is that most of the time is spent intersecting bounding boxes, something that seems to be universally true. It also can tell you how many bounding boxes and primitives are checked.

Like many algorithms, the speed will vary based on the input. Obviously, large data sets tend to take more time than small ones, but the structure of the models you use for benchmarking is also important. Ideally, you use a set of models that are characteristic of the types of models you expect to use.

Profiling is especially critical for low-level optimizations. Intuition is often very wrong about what changes will make the code faster and which ones the compiler was already doing for you. Compilers are good at rearranging nearby instructions. They are bad at knowing that the value you are continually reading through three levels of indirection is constant. Keeping things clean and local makes a big difference. This paper makes almost no attempt to deal with this level of optimization.

2.3. Complexity is Bad

Complexity in the intersection algorithm causes problems in many ways. The more complex your code becomes, the more likely it is to behave unexpectedly on new data sets. Additionally, complexity usually means branching, which is significantly slower than similar code with few branches. If you are checking the state of something in order to avoid extra work, it is important that the amount of work is significant and that you actually avoid the work often enough to justify the checks. This is the argument against the caches used in Section 4.4.

2.4. Preprocessing is Good

In many of the situations where ray casting is used, it is very common to cast hundreds-of-millions of rays. This usually takes a much longer time than it took to build the ray-tracing data structures. A large percentage increase in the time it takes to build the data structures may provide a significant win even if the percentage decrease in the ray-casting time of each ray is much smaller. Ideally, you increase the complexity and sophistication of the hierarchy building stage in order to reduce the complexity and number of intersections computed during the ray traversal stage. This principle motivates Section 4.3.

2.5. *Compute Only What You Need*

There are many different algorithms for many of the components of ray casting. Often there are different algorithms because different information is
needed out of them. Much of the following discussion will be based on the
principle of determining the minimum amount of information needed and then
computing or using that and nothing more. Often this results in a faster algorithm. Examples of this will be shown in Sections 4.1 and 4.2.

3. Overview of Bounding Volume Hierarchies

A bounding-volume hierarchy is simply a tree of bounding volumes. The
bounding volume at a given node encloses the bounding volumes of its children. The bounding volume of a leaf encloses a primitive. If a ray misses the
bounding volume of a particular node, then the ray will miss all of its children, and the children can be skipped. The ray-casting algorithm traverses
this hierarchy, usually in depth-first order, and determines if the ray intersects
an object.

   There are several ways of building bounding-volume hierarchies ([Gold-
smith, Salmon 87], [Rubin, Whitted 80]). The simplest way to build them
is to take a list of bounding volumes containing the primitives and sort along
an axis [Kay, Kajiya 86]. Split the list in half, put a bounding box around
each half, and then recurse, cycling through the axes as you recurse. This is
expressed in pseudocode in Figure 1. This method can be modified in many
ways to produce better hierarchies. A better way to build the hierarchy is
to try to minimize the cost functions described by Goldsmith and Salmon
[Goldsmith, Salmon 87].

```
BoundingVolume BuildHierarchy(bvList, start, end, axis)
   if(end==start)      // only a single bv in list so return it.
        return bvList[start]
   BoundingVolume parent
   foreach bv in bvList
        expand parent to enclose bv
   sort bvList[start..end] along axis
   axis = next axis
   parent.AddChild(BuildHierarchy(bvList, start, (start+end)/2, axis)
   parent.AddChild(BuildHierarchy(bvList, 1+(start+end)/2, end, axis)
   return parent
```

Figure 1. Building a bounding volume hierarchy recursively.

```
bool RaySlabsIntersection(ray, bbox)
   Interval inside = ray.Range()
   for i in (0,1,2)
      inside =
         Intersection(inside, (slab[i].Range()-ray.Origin[i])/ray.Direction[i])
      if(inside.IsEmpty())
         return false
   return true
```

Figure 2. Pseudocode for intersecting a ray with a box represented as axis aligned slabs.

4. Optimizations for Bounding-Volume Hierarchies

4.1. Bounding-Box Intersections

Intersecting rays with bounding volumes usually accounts for most of the time spent casting rays. This makes bounding-volume intersection tests an ideal candidate for optimization. The first issue is what sort of bounding volumes to use. Most of the environments I work with are architectural and have many axis-aligned planar surfaces. This makes axis-aligned bounding boxes ideal. Spheres tend not to work very well for this type of environment.

There are many ways to represent and intersect an axis-aligned bounding box. I have seen bounding box-code that computed the intersection point of the ray with the box. If there was an intersection point, the ray hits the box, and if not, the ray misses. There are optimizations that can be made to this approach, such as making sure you only check faces that are oriented towards the ray, and taking advantage of the fact that the planes are axis aligned [Woo 90]. Still, the approach is too slow. The first hint of this is that the algorithm computes an intersection point. We don't care about that, we just want a yes or no answer. Kay [Kay, Kajiya 86] represented bounding volumes as the intersection of a set of slabs (parallel planes). A slab is stored as a direction D_s, and an interval I_s, representing the minimum and maximum value in that direction, effectively as two plane equations. The set of slab directions is fixed in advance. In my experience, this approach is most effective when there are three, axis aligned, slab directions. This is just another way of storing a bounding box, we store minimum and maximum values along each axis.

Given this representation, we can intersect a bounding box fairly efficiently. We show this in pseudocode in Figure 2. This code isn't as simple as it looks due to the comparisons of the IsEmpty and Intersection functions and the need to reverse the minimum and maximum values of the interval when dividing by a negative number, but it is still much faster than computing the intersection point with the box.

One important thing to notice about this representation and this intersection code is that it gives the right answer when the ray direction is 0 for a particular component. In this case, the ray is parallel to the planes of the slab. Dividing by zero gives either $[-\infty, -\infty]$ or $[+\infty, +\infty]$ when the ray is outside the slab and $[-\infty, +\infty]$ when the ray is inside. This saves additional checks on the ray direction.

4.2. Intersection Rays Versus Shadow Rays

It is important to know what kind of information you need from the ray-casting algorithm in order to keep from doing more work than necessary. There are three commonly used ray-casting queries: closest hit, any hit, and all hits. Closest hit is used to determine the first object in a given direction. This query is usually used for primary, reflected, and transmitted rays. Any hit is used for visibility tests between two points. This is done when checking to see if a point is lit directly by a light and for visibility estimation in radiosity algorithms. The object hit is not needed, only the existence of a hit. All hits is used for evaluating CSG models directly. The CSG operations are performed on the list of intervals returned from the all hits intersection routine.

For efficiency, it is important to keep these queries separate. This can be seen by looking at what happens when using the most general query, all hits, to implement the others. Any hit will simply check to see if the list of intersections is empty. Clearly, we computed more than we needed in this case. Closest hit will sort the list and return the closest intersection. It may seem as if the same or more work is needed for this query; however, this is usually not the case. With most ray-tracing efficiency schemes, once an intersection is found, parts of the environment beyond the intersection point can be ignored. Finding intersections usually speeds up the rest of the traversal. Also, the list of hit data does not need to be maintained.

Shadow (any hit) rays are usually the most common type of rays cast, often accounting for more than 90 percent of all rays. Therefore, it is worth considering how to make them faster than other types of rays. Shadow rays need not compute any of the commonly needed intersection information, such as intersection point, surface normal, uv coordinates, or exact object hit. Additionally, the traversal of the efficiency structure can be terminated immediately once an intersection is guaranteed. A special shadow routine taking these factors into account can make a significant difference in efficiency.

The difference between shadow rays and intersection rays determine which acceleration scheme I use. I have tried both grids [Fujimoto et al. 86] and bounding-volume hierarchies. In my experience (based on models I typically render), grids are a little faster on intersection rays (closest hit) and slower for shadow rays (any hit). Grids sort the environment spatially, which is good

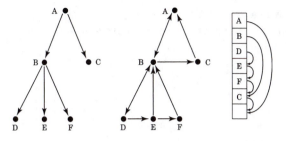

Figure 3. Three different representations for a tree. (a) Children pointers. (b) Left child, right sibling, parent pointers. (c) Array in depth-first order with skip pointers.

for finding the closest intersection. The bounding-volume hierarchies built by trying to minimize Goldsmith and Salmon's cost function [Goldsmith, Salmon 87] tend to keep larger primitives near the root, which is good for shadow rays. It is still unknown which acceleration scheme is better; it is almost certainly based on the model.

4.3. Traversal Code

Casting a ray against a bounding-volume hierarchy requires traversing the hierarchy. If a ray hits a bounding volume, then the ray is checked against the children of the bounding volume. If the bounding volume is a leaf, then it has an object inside it, and the object is checked. This is done in depth-first order. Once bounding-volume intersection tests are as fast as they can be, the next place for improvement is the traversal of the hierarchy. Traversal code for shadow rays will be used in the following discussion.

In 1991, Haines [Haines 91] published some techniques for better traversals. Several of these techniques used extra knowledge to mark bounding boxes as automatically hit and to change the order of traversal. In my experience, these methods do not speed up the ray tracer and greatly increase the complexity of the code. This difference in experience may be due to changes in architecture over the last eight years making branches and memory accesses the bottleneck, rather than floating point operations. The differing result may also be due to faster bounding-box tests. I have found that the best way to make the traversal fast is to make it as minimal as possible.

The simplest traversal code is to use recursion to traverse the tree in depth-first order. Figure 3(a) shows a hierarchy of bounding boxes. Depth first traversal means that bounding box A is tested, then box B, then the boxes with primitives D, E, and F, and finally box C. The idea is to find an inter-

173

TreeShadowTraversal(ray, bvNode)
  while (true)   // *terminates when bvNode→GetParent() is*   // *NULL*
    if (bvNode→Intersect(ray))
      if (bvNode→HasPrimitive())
        if (bvNode→Primitive().Intersect(ray))
          return true
    else
      bvNode = bvNode→GetLeftChild()
      continue
    while(true)
      if(bvNode→GetRightSibling() != NULL)
        bvNode = bvNode→GetRightSibling()
        break
      bvNode = bvNode→GetParent()
      if(bvNode == NULL)
        return false

Figure 4. Traversal of bounding volume tree using left child, right sibling, parent structure.

section as soon as possible by traveling down the tree. The biggest problem with this is the function-call overhead. The compiler maintains much more state information than we need here. We can eliminate much of this overhead by changing our representation of the tree. A representation that works well is to store the left-most child, the right sibling, and the parent for each node, as in Figure 3. Using this representation, we can get rid of the recursion by following the appropriate pointers. If the ray intersects the bounding box, we get to its children by following the left-most child link. If the ray misses, we get to the next node by following the right sibling link. If the right sibling is empty, we move up until either there is a right sibling, or we get back up to the root, as shown in pseudocode in Figure 4.

This tree traversal also does too much work. Notice that when the traversal is at a leaf or when the ray misses a bounding volume, we compute the next node. The next node is always the same; there is no reason to be computing it for each traversal. We can pre-compute the node we go to when we skip this subtree and store this skip node in each node. This step eliminates all computation of traversal-related data from the traversal. There are still intersection computations, but no extra computation for determining where to go. This is expressed in pseudocode in Figure 5.

The final optimization is the recognition that we only need to do depth-first traversals on the tree once it is built. This observation lets us store the tree in an array in depth-first order as in Figure 3. If the bounding volume is intersected, the next node to try is the next node in the array. If the bounding

```
SkipTreeShadowTraversal(ray, bvNode)
  while(bvNode != NULL)
    if(bvNode→Intersect(ray))
      if(bvNode→HasPrimitive())
        if(bvNode→Primitive().Intersect(ray))
          return true
        bvNode = bvNode→SkipNode()
      else
        bvNode = bvNode→GetLeftChild()
    else
      bvNode = bvNode→SkipNode()
  return false
```

Figure 5. Traversal of bounding volume tree using left child and skip pointers.

volume is missed, the next node can be found through the skip mechanism. We have effectively eliminated all the information we don't need from the tree, yet it is still possible to reconstruct it. The traversal code can be seen in Figure 6.

The array-traversal approach works significantly better than the previous one, and has some subtle advantages. The first is better memory usage. In addition to the bounding volume, this method requires only a pointer to a primitive and a pointer to the skip node. This is very minimal. Since the nodes are arranged in the order they will be accessed, there is more memory coherency for large environments. The second advantage is that this method requires copying data from the original tree into an array. Since the original tree is going to be discarded, it can be augmented with extra information. Depending upon how the tree is created, this extra information can more

```
ArrayShadowTraversal(ray, bvNode)
  stopNode = bvNode→GetSkipNode()
  while(bvNode < stopNode)
    if(bvNode→Intersect(ray))
      if(bvNode→HasPrimitive())
        if(bvNode→Primitive().Intersect(ray))
          return true
      bvNode++
    else
      bvNode = bvNode→GetSkipNode()
  return false
```

Figure 6. Traversal of bounding volume tree stored as an array in depth-first order.

than double the cost of each node. With our method, there is no penalty for this information. Storing the extra information can reduce the time to build the tree and more importantly, can result in better trees. The fastest bounding-volume test is the one you don't have to do.

4.4. Caching Objects

One common optimization is the use of caches for the object most recently hit. This optimization, and variations on it, are discussed by Haines [Haines 91]. The idea is that the next ray cast will be similar to the current ray, so keep the intersected object around and check it first the next time. To the extent that this is true, caches can provide a benefit; however, rays often differ wildly. Also, cache effectiveness decreases as the size of the primitives get smaller. The realism of many types of models is increased by replacing single surfaces with many surfaces. As a result, caches will remain valid for a shorter amount of time.

There are two different types of caches, those for intersection (closest hit) rays and those for shadow (any hit) rays. If caches are used for intersection rays, the ray will still need to be checked against the environment to see if another object is closer. Usually the ray will again be checked against whatever object is in the cache. Mailboxes [Arnaldi et al. 87] can eliminate this second check (by marking each tested object with a unique ray ID and then checking the ID before testing the primitive). Mailboxes, however, create problems when making a parallel version of the code. Depending on the environment and the average number of possible hits-per-ray, the cache may reduce the size of the environment that must be checked by shortening the ray length. In my experience, the cost of maintaining the cache and the double intersection against an object in it more than outweighs the benefit of having a cache. If your primitives are very expensive and your environments are dense, the benefit of reducing the length of the ray early may outweigh the costs, but it is worth checking carefully.

Evaluating the benefit of caches for shadow rays is more complicated. In cases where there is a single light, there tends to be a speed-up as long as the cache remains full much of the time and the objects in it stay there for a long enough time. In cases where there are multiple lights, we often lose shadow-ray coherence because the lights are in different regions of the environment. Since each shadow ray is significantly different from the previous one, the solution is to have a different cache for each light.

For both types of caches, we have ignored what happens for reflected and transmitted rays. These rays are spatially very different from primary rays and from each other. Each additional bounce makes the problem much worse. If rays are allowed to bounce d times, there are $2^{d+1} - 1$ different nodes in the ray tree. In order for caching to be useful, a separate cache needs to be associated with each node. For shadow rays, that means a separate cache for each light at each node This can increase the complexity of the code significantly. Another option is to store a cache for each light on each object (or collection of objects) in the environment as discussed by Haines [Haines 91]. Note that caching only helps when there is an object in the cache. If most shadow rays won't hit anything (due to the model or the type of algorithm using the shadow tests), then the cache is less likely to be beneficial. In my experience, shadow caching wasn't a significant enough gain, so I opted for simplicity of code and removed it; although, after generating the data for the result section, I am considering putting it back in for certain situations. Others have found that caches are still beneficial.

5. Results

Now we look at the cumulative effects for shadow rays of the three main optimizations described in the paper. First, we speed-up bounding-box tests. Next, we speed-up the traversal using the different methods from Section 4.3. We then treat shadow rays differently from intersection rays and lastly we add a shadow cache. In all of the experiments, 1,000,000 rays are generated by choosing random pairs of points from within a bounding box 20% larger than the bounding box of the environment. In the last experiment, 500,000 rays are generated, each generated ray is cast twice, resulting in 1,000,000 rays being cast overall. The first two test cases are real environments, the rest are composed of randomly-oriented and -positioned unit right triangles. The number gives the number of triangles. Small, mid, and big refer to the space the triangles fill. Small environments are 20-units-cubed, mid are 100-units-cubed, and big are 200-units-cubed. The theater model has 46,502 polygons. The science center model has 4,045 polygons. The code was run on an SGI O2 with a 180 MHz R5000 using the SGI compiler with full optimization.[1] No shading or other computation was done and time to build the hierarchies was not included.

The experiments reported in Table 1 are explained in more detail below:

1. Bounding box test computes intersection point, traversal uses recursion, and shadow rays are treated as intersection rays.

2. Bounding box test replaced by slab version from Section 4.1.

3. Recursive traversal replaced by iterative traversal using left child, right sibling, and parent pointers as in Section 4.3.

---

[1]-Ofast=ip32_5k

177

	1	2	3	4	5	6	7	8
theater	64	36	30	21	22	11	10	6
lab	79	41	32	22	20	12	12	7
10,000 small	415	223	191	142	110	48	50	27
10,000 mid	392	185	154	103	81	77	79	65
10,000 big	381	179	152	104	82	79	77	69
100,000 small	995	620	550	449	351	62	63	33
100,000 mid	932	473	424	324	230	146	148	89
100,000 big	1024	508	442	332	240	210	212	156
300,000 mid	1093	597	536	421	312	120	121	64

Table 1. Results of the different experiments described in the text on different environments. Times are rounded to the nearest second.

4. Skip pointer used to speed up traversal as in Section 4.3.

5. Tree traversal replaced by array traversal as in Section 4.3.

6. Intersection rays replaced by shadow rays as in Section 4.2.

7. Shadow caching used as in Section 4.4.

8. Shadow caching used, but each ray checked twice before generating a new ray. The same number of checks were performed.

The first thing to notice is that real models require much less work than random polygons. This is because the polygons are distributed very unevenly and vary greatly in size. The theater has a lot more open space and even more variation in polygon size than the lab, resulting in many inexpensive rays and a faster average time. In spite of this, the results show very similar trends for all models. In the first five experiments we have not used any model-specific knowledge; we have just reduced the amount of work done. Special shadow rays and caching are more model specific. Shadow rays are more effective when there are many intersections along the ray and are almost the same when there is zero or one intersection. Shadow caching is based on ray coherence and the likelihood of having an intersection. In Experiment 7 there is an unrealistically low amount of coherence (none). In Experiment 8 we guaranteed that there would be significant coherence by casting each ray twice.

6. Conclusions

The optimization of ray-casting code is a double-edged sword. With careful profiling it can result in significant speed-ups. It can also lead to code that is slower and more complicated. The optimizations presented here are probably fairly independent of the computer architecture. There are plenty of significant, lower-level optimizations that can be made which may be completely dependent upon the specific platform. If you plan on porting your code to other architectures, or even keeping your code for long enough that the architecture changes under you, these sorts of optimizations should be made with care.

Eventually, you reach a point where further optimization makes no significant difference. At this point, you will have no choice but to go back and try to create better trees requiring fewer primitive and bounding box tests, or to look at entirely different acceleration strategies. Over time, the biggest wins come from better algorithms, not better code tuning.

The results presented here should be viewed as a case study. They describe some of what has worked for me on the types of models I use. They may not be appropriate for the types of models you use.

Acknowledgments. Thanks to Peter Shirley, Jim Arvo, and Eric Haines for many long discussions on ray tracing. Thanks to Peter and Eric for encouraging me to write up these experiences, and to both of them and Bill Martin for helpful comments on the paper. This work was partially funded by Honda and NSF grant ACI-97-20192.

References

[Arnaldi et al. 87] B. Arnaldi, T. Priol, and K. Bouatouch. "A New Space Subdivision Method for Ray Tracing CSG Modelled Scenes." *The Visual Computer*, 3(2):98–108 (August 1987).

[Bentley 82] J. L. Bentley. *Programming Pearls (reprinted with corrections)*. Reading, MA: Addison-Wesley, 1989.

[Bentley 89] J. L. Bentley. *Writing Efficinet Programs*. Englewood Cliffs, NJ: Prentice-Hall, 1982.

[Fujimoto et al. 86] A. Fujimoto, T. Tanaka, and K. Iwata. "Arts: Accelerated Ray-tracing System." *IEEE Computer Graphics and Applications* :16–20 (April 1986).

[Glassner 89] A. Glassner, editor. *An Introduction to Ray Tracing*. London: Academic Press, 1989.

Vol. 3, No. 2: 1–14

[Goldsmith, Salmon 87] J. Goldsmith and J. Salmon. "Automatic Creation of Object Hierarchies for Ray Tracing." *IEEE Computer Graphics and Applications* 7(5):14–20 (May 1987).

[Haines 91] E. Haines. "Efficiency Improvements for Hierachy Traversal." In *Graphics Gems II,* edited by James Arvo, pp. 267–273. San Diego: Academic Press, 1991.

[Kay, Kajiya 86] T. L. Kay and J. T. Kajiya. "Ray Tracing Complex Scenes." In *Computer Graphics (Proc. SIGGRAPH 86),* edited by D. C. Evans and R. J. Athay, pp. 269–278, 1986.

[Knuth 92] D. E. Knuth. *Literate Programming, CSLI Lecture Notes Number 27.* Stanford, CA: Stanford University Center for the Study of Language and Information, 1992.

[Rubin, Whitted 80] S. M. Rubin and T. Whitted. "A 3-Dimensional Representation for Fast Rendering of Complex Scenes." *Computer Graphics* 14(3):110–116 (July 1980)

[Woo 90] Andrew Woo. "Fast Ray-Box Intersection." In *Graphics Gems,* edited by A. Glassner, pp. 395–396. San Diego: Academic Press, 1990.

Web Information:

Additional information is available at
http://www.acm.org/jgt/papers/Smits98

Brian Smits, University of Utah, Department of Computer Science, 50 S. Central Campus Drive, RM 3190, Salt Lake City, UT 84112-9205 (bes@cs.utah.edu)

Received October 16, 1998; accepted January 18, 1999.

Current Contact Information:

Brian Smits, Pixar Animation Studios, 1200 Park Ave, Emeryville, CA 94608 (bes@pixar.com)

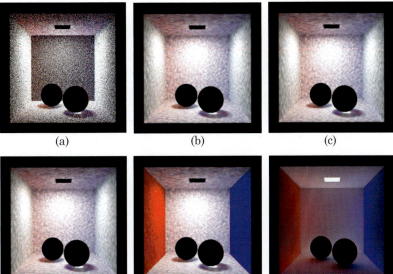

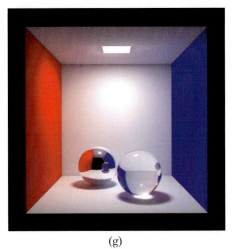

Plate I. Cornell box with spheres: (a) Global photon map. (b) Irradiance estimates computed at image sample points (dimmed for display). (c) Precomputed irradiance estimates at all 400,000 photon positions (dimmed for display). (d) Precomputed irradiance estimates at 100,000 photon positions (dimmed for display). (e) Radiance estimates based on (d). (f) Soft indirect illumination computed with final gathering. (g) Complete image with direct illumination, specular reflection and refraction, caustics, and soft indirect illumination. (See Figure 1, page 246.)

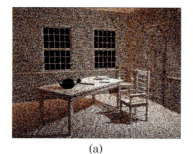

(a)   (b)

(c)   (d)

(e)

Plate II. Interior: (a) Global photon map. (b) Irradiance estimates computed precisely at the image sample points (dimmed for display). (c) Precomputed irradiance estimates at all 500,000 photon positions (dimmed for display). (d) Radiance estimates based on (c). (e) Complete image with direct illumination, specular reflection, and soft indirect illumination. (See Figure 2, page 248.)

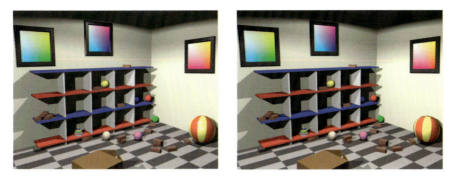

Plate III. Rendering of direct illumination using RGB renderer (left) and spectral renderer (right). (See Figure 6, page 299.)

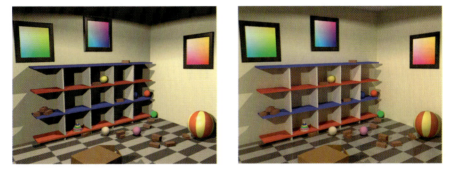

Plate IV. Rendering with nonwhite light sources using RGB renderer with direct lighting (left) and spectral render with direct and indirect illumination (right). (See Figure 7, page 300.)

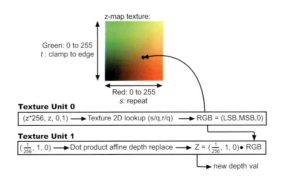

Plate V. Texture shader setup. (See Figure 4, page 221.)

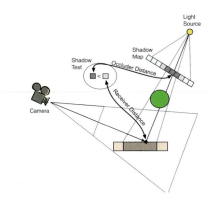

Plate VI. Shadow mapping illustrated. (See Figure 1, page 218.)

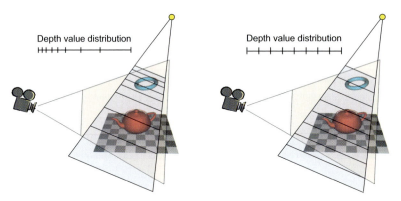

Plate VII. Distribution of depth values. Left: $1/z$. Right: z linear. (See Figure 2, page 219.)

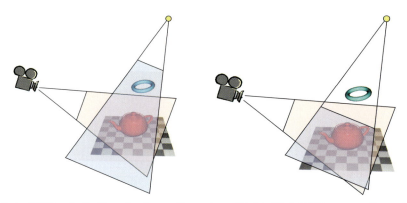

Plate VIII. Left: Standard near/far setup. Right: Tight-fitting near/far setup. (See Figure 3, page 220.)

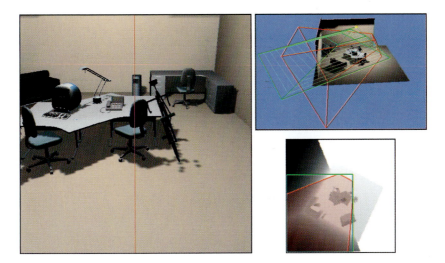

Plate IX. Left: Scene rendered with light frustum adjusted (left half) compared to the traditional method (right half). Top right: The optimized light frustum (green) and camera frustum (red). Bottom right: Location of convex hull (red) and minimum area enclosing rectangle (green) in the nonoptimized shadow map. (See Figure 6, page 224.)

Plate X. Left: Camera close-up (left half with adjusted light frustum, right half without). Top right: Optimized light frustum (green) and camera frustum (red). Bottom right: nonoptimized shadow map and optimized shadow map. (See Figure 7, page 225.)

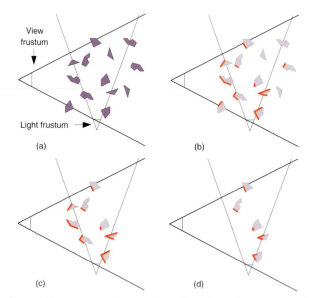

(a)

(b)

(c)

(d)

Plate XI. A two-dimensional illustration of regions that have to be processed when shadow maps are used. (a) A view frustum, a light frustum, and several objects. (b) The regions that are visible from the viewpoint are marked with bold red. (c) The visible regions that belong to objects that intersect with the light frustum. (d) The visible regions that are inside the light frustum and could thus be affected by the light source. Our algorithm processes only the regions marked in (d). (See Figure 1, page 231.)

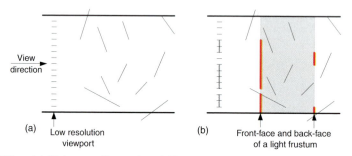

(a) Low resolution viewport

(b) Front-face and back-face of a light frustum

Plate XII. (a) This two-dimensional illustration shows a low-resolution viewport and several opaque objects (lines) in post-perspective space. The dark lines indicate the view frustum. (b) The visible parts of the front-face and the back-face of a light frustum are marked with bold red. The light source can affect only the parts of the objects that are inside the gray light frustum. In the visible subset of these parts, the front-face of the light frustum is visible (bold red) and the corresponding back-face is hidden. In practice, the test can be implemented using a stencil buffer. (See Figure 2, page 232.)

Plate XIII. The flashlight scene consists of 77K triangles and one spotlight. The room scene uses the same geometry, illuminated by one omnidirectional light source, implemented using six spotlights without the angular attenuation ramp. The parking lot scene has 101K triangles and 12 spotlights. (See Figure 4, page 235.)

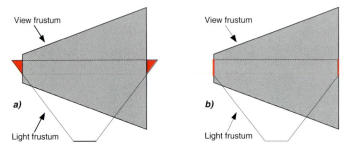

Plate XIV. (a) A two-dimensional illustration of the light frustum intersecting with the near and far clipping planes of the view frustum. (b) The light frustum has been clipped by the near and far planes of the view frustum. The light frustum must remain closed after clipping, and thus the parts outside the near and far planes (marked with red) need to be projected to the planes by using parallel projection. (See Figure 3, page 233.)

Plate XV. Glass with nested dielectrics. Priorities (from highest to lowest) are: the glass, the ice cube, the air bubbles under the water, and the water. (See Figure 3, page 197.)

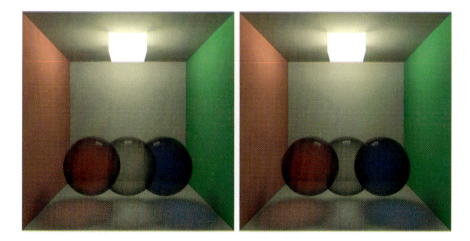

Plate XVI. The same scene rendered with different priorities. In the first image, priorities decrease from left to right. In the second, the middle sphere has the lowest priority. To make the difference more visible, the spheres have been given the same refractive indices as the surrounding air. (See Figure 1, page 192.)

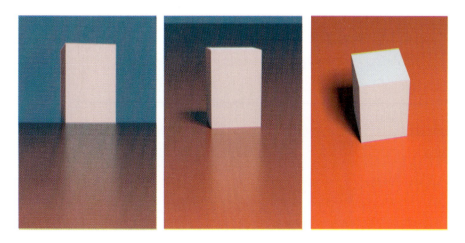

Plate XVII. Three views for $n_u = n_v = 400$ and a red substrate. (See Figure 3, page 306.)

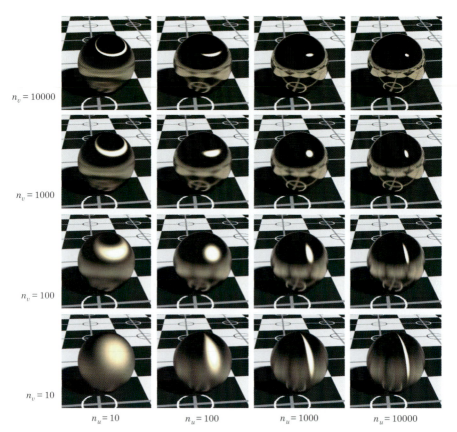

$n_v = 10000$

$n_v = 1000$

$n_v = 100$

$n_v = 10$

$n_u = 10$　　$n_u = 100$　　$n_u = 1000$　　$n_u = 10000$

Plate XVIII. Metallic spheres for various exponents. (See Figure 1, page 305.)

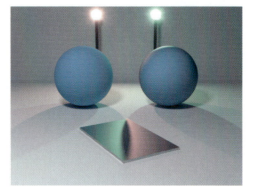

Plate XIX. An image with a Lambertian sphere (left) and a sphere with $n_u = n_v = 5$. After a figure from Lafortune et al. [Lafortune et al. 97]. (See Figure 4, page 306)

Plate XX. A closeup of the model implemented in a path tracer with 9, 26, and 100 samples. (See Figure 6, page 309.)

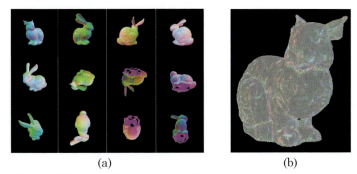

(a)   (b)

Plate XXI. (a) The 12 different views of an object used during normal map creation for the bunny model. The object is rendered with colors representing the vertex normal vectors. (b) The normal errors for the bunny model corresponding to a given view. (See Figure 2, page 97.)

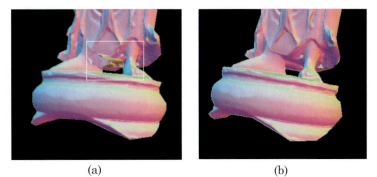

(a)   (b)

Plate XXII. Comparison of the methods without and with normal initialization. Both images are synthesized by rendering a simplified model (Buddha) with normal map textures. The left image corresponds to the method without normal initialization, while the right side corresponds to the method with normal initialization. The white rectangle contains some invisible parts for the fixed set of viewpoints, which are below the robe of the statue. (See Figure 3, page 99.)

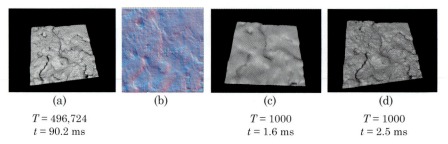

(a)   (b)   (c)   (d)

$T = 496{,}724$ $T = 1000$ $T = 1000$
$t = 90.2$ ms $t = 1.6$ ms $t = 2.5$ ms

Plate XXIII. Normal maps for height fields mesh can be generated from a single view point. T is the number of triangles of the model; t is the drawing time per frame. (b) Normals are color-coded. (c) Rendering of the simplified model using Gouraud shading on a PC with a Geforce 3 graphics card. (d) Rendering of the simplified model with a normal map on the same PC as (c). (See Figure 1, page 96.)

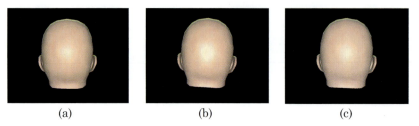

(a)   (b)   (c)

Plate XXIV. Results for different internal and external normal map precisions. (a) The external normal map type is GL_SIGNED_BYTE; the internal format is GL_SIGNED_HILO_NV. (b) The external normal map type is GL_RGB8; the internal format is the same. (c) The external normal map type is GL_SHORT; the internal format is GL_SIGNED_HILO_NV. (See Figure 4, page 99.)

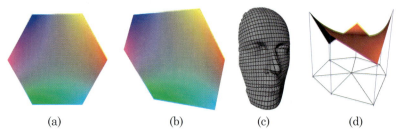

(a)   (b)   (c)   (d)

Plate XXV. (a) Smooth color blending using barycentric coordinates for regular polygons [Loop, DeRose 89]. (b) Smooth color blending using our generalization to arbitrary polygons. (c) Smooth parameterization of an arbitrary mesh using our new formula which ensures nonnegative coefficients. (d) Smooth position interpolation over an arbitrary convex polygon (S-patch of depth 1). (See Figure 1, page 60.)

Plate XXVI. Original image. (See Figure 3, page 322.)

Plate XXVII. Day-for-night tone mapping. (See Figure 4, page 322.)

Plate XXVIII. Blurred to remove fine detail. (See Figure 5, page 323.)

Plate XXIX. Night-filtered, with the same level of fine detail as in Plate XXVIII. (See Figure 6, page 323.)

Plate XXX. Blurred plus noise. (See Figure 7, page 324.)

Plate XXXI. Night filtering plus noise. (See Figure 8, page 324.)

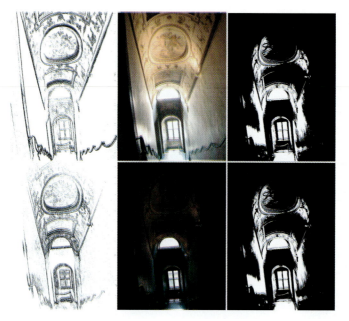

Plate XXXII. Two unaligned exposures (middle) and their corresponding edge bitmaps (left) and median threshold bitmaps (right). The edge bitmaps are not used in our algorithm, precisely because of their tendency to shift dramatically from one exposure level to another. In contrast, the MTB is stable with respect to exposure. (See Figure 1, page 331.)

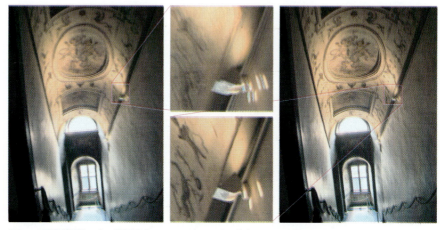

Plate XXXIII. An HDR image composited from unaligned exposures (left) and detail (top center). Exposures aligned with our algorithm yield a superior composite (right) with clear details (bottom center). (See Figure 6, page 341.)

Vol. 2, No. 1: 21–28

Fast, Minimum Storage Ray-Triangle Intersection

Tomas Möller
Prosolvia Clarus AB

Ben Trumbore
Cornell University

Abstract. We present a clean algorithm for determining whether a ray intersects a triangle. The algorithm translates the origin of the ray and then changes the base to yield a vector $(t\ u\ v)^T$, where t is the distance to the plane in which the triangle lies and (u, v) represents the coordinates inside the triangle.

One advantage of this method is that the plane equation need not be computed on the fly nor be stored, which can amount to significant memory savings for triangle meshes. As we found our method to be comparable in speed to previous methods, we believe it is the fastest ray-triangle intersection routine for triangles that do not have precomputed plane equations.

1. Introduction

A ray $R(t)$ with origin O and normalized direction D is defined as

$$R(t) = O + tD, \tag{1}$$

and a triangle is defined by three vertices V_0, V_1, and V_2. In the ray-triangle intersection problem we want to determine if the ray intersects the triangle.

Previous algorithms have solved this problem by first computing the intersection between the ray and the plane in which the triangle lies and then testing if the intersection point is inside the edges [Haines 94].

Our algorithm uses minimal storage (i.e., only the vertices of the triangle need to be stored) and does not need any preprocessing. For triangle meshes, the memory savings are significant, ranging from about 25 percent to 50 percent, depending on the amount of vertex sharing.

In our algorithm, a transformation is constructed and applied to the origin of the ray. The transformation yields a vector containing the distance t to the intersection and the coordinates (u, v) of the intersection. In this way the ray-plane intersections of previous algorithms are avoided. Note that this method has been discussed previously, by for example Patel and Shirley [Patel 96] [Shirley 96].

2.   Intersection Algorithm

A point $T(u, v)$ on a triangle is given by

$$T(u, v) = (1 - u - v)V_0 + uV_1 + vV_2, \tag{2}$$

where (u, v) are the barycentric coordinates, which must fulfill $u \geq 0$, $v \geq 0$, and $u + v \leq 1$. Note that (u, v) can be used for texture mapping, normal interpolation, color interpolation, and more. Computing the intersection between the ray $R(t)$ and the triangle $T(u, v)$ is equivalent to $R(t) = T(u, v)$ which yields:

$$O + tD = (1 - u - v)V_0 + uV_1 + vV_2. \tag{3}$$

Rearranging the terms gives:

$$\begin{bmatrix} -D, & V_1 - V_0, & V_2 - V_0 \end{bmatrix} \begin{bmatrix} t \\ u \\ v \end{bmatrix} = O - V_0. \tag{4}$$

This result means the barycentric coordinates (u, v) and the distance t from the ray origin to the intersection point can be found by solving the linear system of equations above.

The above can be thought of geometrically as translating the triangle to the origin and transforming it to a unit triangle in y and z with the ray direction aligned with x, as illustrated in Figure 1 (where $M = [-D, V_1 - V_0, V_2 - V_0]$ is the matrix in Equation (4)).

Arenberg [Arenberg 88] describes a similar algorithm to the one above. He also constructs a 3×3 matrix but uses the normal of the triangle instead of the ray direction D. This method requires storing the normal for each triangle or computing it on the fly.

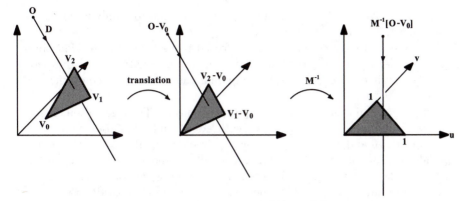

Figure 1. Translation and change of base of the ray origin.

Denoting $E_1 = V_1 - V_0$, $E_2 = V_2 - V_0$, and $T = O - V_0$, the solution to Equation (4) is obtained by using Cramer's rule:

$$
\begin{bmatrix} t \\ u \\ v \end{bmatrix} = \frac{1}{|-D,\ E_1,\ E_2|} \begin{bmatrix} |\ T,\ E_1,\ E_2\ | \\ |-D,\ T,\ E_2| \\ |-D,\ E_1,\ T| \end{bmatrix}. \tag{5}
$$

From linear algebra, we know that $|A,\ B,\ C| = (A \times B) \cdot C = -(A \times C) \cdot B = -(C \times B) \cdot A$. Equation (5) could therefore be rewritten as

$$
\begin{bmatrix} t \\ u \\ v \end{bmatrix} = \frac{1}{(D \times E_2) \cdot E_1} \begin{bmatrix} (T \times E_1) \cdot E_2 \\ (D \times E_2) \cdot T \\ (T \times E_1) \cdot D \end{bmatrix} = \frac{1}{P \cdot E_1} \begin{bmatrix} Q \cdot E_2 \\ P \cdot T \\ Q \cdot D \end{bmatrix}, \tag{6}
$$

where $P = (D \times E_2)$ and $Q = (T \times E_1)$. In our implementation we reuse these factors to speed up the computations.

3. Implementation

The following C implementation (available online) has been tailored for optimum performance. Two branches exist in the code; one which efficiently culls all back-facing triangles (#ifdef TEST_CULL) and the other which performs the intersection test on two-sided triangles (#else). All computations are delayed until they are required. For example, the value for v is not computed until the value of u is found to be within the allowable range.

The one-sided intersection routine eliminates all triangles where the value of the determinant (det) is negative. This procedure allows the routine's only division operation to be delayed until an intersection has been confirmed. For shadow test rays this division is not needed at all, since our only requirement is whether the triangle is intersected.

The two-sided intersection routine is forced to perform this division operation in order to evaluate the values of u and v. Alternatively, this function could be rewritten to conditionally compare u and v to zero based on the sign of det.

Some aspects of this code deserve special attention. The calculation of edge vectors can be done as a preprocess, with edge1 and edge2 being stored in place of vert1 and vert2. This speedup is only possible when the actual spatial locations of vert1 and vert2 are not needed for other calculations and when the vertex location data is not shared between triangles.

To ensure numerical stability, the test which eliminates parallel rays must compare the determinant to a small interval around zero. With a properly adjusted EPSILON value, this algorithm is extremely stable. If only front-facing triangles are to be tested, the determinant can be compared to EPSILON, rather than zero (a negative determinant indicates a back-facing triangle).

The value of u is compared to an edge of the triangle ($u = 0$) and also to a line parallel to that edge but passing through the opposite point of the triangle ($u = 1$). Although not actually testing an edge of the triangle, this second test efficiently rules out many intersection points without further calculation.

```
#define EPSILON 0.000001
#define CROSS(dest,v1,v2) \
          dest[0]=v1[1]*v2[2]-v1[2]*v2[1]; \
          dest[1]=v1[2]*v2[0]-v1[0]*v2[2]; \
          dest[2]=v1[0]*v2[1]-v1[1]*v2[0];
#define DOT(v1,v2) (v1[0]*v2[0]+v1[1]*v2[1]+v1[2]*v2[2])
#define SUB(dest,v1,v2)
          dest[0]=v1[0]-v2[0]; \
          dest[1]=v1[1]-v2[1]; \
          dest[2]=v1[2]-v2[2];


int
intersect_triangle(double orig[3], double dir[3],
                   double vert0[3], double vert1[3], double vert2[3],
                   double *t, double *u, double *v)
{
   double edge1[3], edge2[3], tvec[3], pvec[3], qvec[3];
   double det,inv_det;

   /* find vectors for two edges sharing vert0 */
   SUB(edge1, vert1, vert0);
   SUB(edge2, vert2, vert0);

   /* begin calculating determinant-also used to calculate U parameter */
```

```
    CROSS(pvec, dir, edge2);

    /* if determinant is near zero, ray lies in plane of triangle */
    det = DOT(edge1, pvec);

#ifdef TEST_CULL           /* define TEST_CULL if culling is desired */
    if (det < EPSILON)
       return 0;

    /* calculate distance from vert0 to ray origin */
    SUB(tvec, orig, vert0);

    /* calculate U parameter and test bounds */
    *u = DOT(tvec, pvec);
    if (*u < 0.0 || *u > det)
       return 0;

    /* prepare to test V parameter */
    CROSS(qvec, tvec, edge1);

     /* calculate V parameter and test bounds */
    *v = DOT(dir, qvec);
    if (*v < 0.0 || *u + *v > det)
       return 0;

    /* calculate t, scale parameters, ray intersects triangle */
    *t = DOT(edge2, qvec);
    inv_det = 1.0 / det;
    *t *= inv_det;
    *u *= inv_det;
    *v *= inv_det;
#else                      /* the non-culling branch */
    if (det > -EPSILON && det < EPSILON)
      return 0;
    inv_det = 1.0 / det;

    /* calculate distance from vert0 to ray origin */
    SUB(tvec, orig, vert0);

    /* calculate U parameter and test bounds */
    *u = DOT(tvec, pvec) * inv_det;
    if (*u < 0.0 || *u > 1.0)
      return 0;

    /* prepare to test V parameter */
    CROSS(qvec, tvec, edge1);
```

```
    /* calculate V parameter and test bounds */
    *v = DOT(dir, qvec) * inv_det;
    if (*v < 0.0 || *u + *v > 1.0)
      return 0;

    /* calculate t, ray intersects triangle */
    *t = DOT(edge2, qvec) * inv_det;
#endif
    return 1;
}
```

4. Results

Badouel [Badouel 90] presented a ray-triangle intersection routine that also computes the barycentric coordinates. We compared this method to ours. The two nonculling methods were implemented in an efficient ray tracer. Figure 2 presents ray-tracing runtimes from a Hewlett-Packard 9000/735 workstation for the three models shown in Figures 3–5. In this particular implementation, the performance of the two methods is roughly comparable (detailed statistics are available online).

Model	Objects	Polygons	Lights	Our method sec.	Badouel sec.
Car	497	83408	1	365	413
Mandala	1281	91743	2	242	244
Fallingwater	4072	182166	15	3143	3184

Figure 2. Contents and runtimes for data sets in Figures 3–5.

5. Conclusions

We present an algorithm for ray-triangle intersection which we show to be comparable in speed to previous methods while significantly reducing memory storage costs, by avoiding storing triangle plane equations.

Acknowledgments.   Thanks to Peter Shirley, Eric Lafortune, and the anonymous reviewer whose suggestions greatly improved this paper.

 This work was supported by the NSF/ARPA Science and Technology Center for Computer Graphics and Scientific Visualization (ASC-8920219), and by the Hewlett-Packard Corporation and by Prosolvia Clarus AB.

Figure 3. Falling Water.

Figure 4. Mandala.

Figure 5. Car (model is courtesy of Nya Perspektiv Design AB).

References

[Arenberg 88] Jeff Arenberg. "Re: Ray/Triangle Intersection with Barycentric Co-ordinates." *Ray Tracing News* 1(11):(November 4, 1988). http://www.acm.org/tog/resources/RTNews/.

[Badouel 90] Didier Badouel. "An Efficient Ray-Polygon Intersection." In *Graphics Gems*, edited by Andrew S. Glassner, pp. 390–393. Boston: Academic Press Inc., 1990.

[Haines 94] Eric Haines. "Point in Polygon Strategies." In *Graphics Gems IV*, edited by Paul S. Heckbert, pp. 24–46. Boston: AP Professional, 1994.

[Patel 96] Edward Patel. Personal communication, 1996.

[Shirley 96] Peter Shirley. Personal communication, 1996.

Vol. 2, No. 1: 21–28   journal of graphics tools

Web Information:

Source code, statistical analysis, and images are available online at
http://www.acm.org/jgt/papers/MollerTrumbore97/

Tomas Möller, Prosolvia Clarus AB, Chalmers University of Technology, Gårdavägen,
S-41250, Gothenberg, Sweden (tompa@clarus.se)

Ben Trumbore, Program of Computer Graphics, Cornell University, Ithaca, NY
14853 (wbt@graphics.cornell.edu)

Received June 17, 1996; accepted October 17, 1996

New Since Original Publication

The terms in Equation 6 in our original paper can be rearranged as shown
below:

$$
\begin{pmatrix} t \\ u \\ v \end{pmatrix} = \frac{1}{-(E_1 \times E_2) \cdot D} \begin{pmatrix} (E_1 \times E_2) \cdot T \\ (T \times D) \cdot E_2 \\ -(T \times D) \cdot E_1 \end{pmatrix} = \frac{1}{-N \cdot D} \begin{pmatrix} N \cdot T \\ R \cdot E_2 \\ -R \cdot E_1 \end{pmatrix}, \quad (7)
$$

where $R = T \times D$, and $N = E_1 \times E_2$ is the unnormalized normal of the
triangle, which is constant if the triangle remains static. If V_0, E_1, E_2, and N
are stored with each triangle, then the equation above can be used to compute
the intersection quicker, since one entire cross product (N) has been lifted out,
precomputed, and stored with the triangle. Compared to the fastest version of
our old algorithm, we have observed about 40% speedup in a simple ray tracer
using this strategy. It should be noted, however, that this defies the original
intent of using minimum storage for computing the ray-triangle intersection.
However, we still hope that some readers might find this optimization useful.

   It should also be noted that our algorithm treats an edge shared by two
triangles differently, depending on which triangle is being intersected. Therefore, it may happen that a ray may miss both triangles if the ray is extremely
close to that shared edge. It can become a bit more robust by adding some
epsilon to the test against u, v, and $u + v$. However, to fully avoid this problem shared edges must be treated in exactly the same way. Plücker-based
algorithms work much better in this sense.

Acknowledgments. Thanks to Henrik Wann Jensen, who also, independently,
came up with this rearrangement.

Current Contact Information:

Tomas Akenine-Möller, Department of Computer Science, Lund University, Box 118, 221 00 Lund, Sweden (tam@cs.lth.se)

Ben Trumbore, Autodesk, Inc., 10 Brown Rd, Suite 2000, Ithaca, NY 14850 (ben@trumbore.com)

Vol. 7, No. 2: 1–8

Simple Nested Dielectrics in Ray Traced Images

Charles M. Schmidt and Brian Budge

University of Utah

Abstract. This paper presents a simple method for modeling and rendering refractive objects that are nested within each other. The technique allows the use of simpler scene geometry and can even improve rendering time in some images. The algorithm can be easily added into an existing ray tracer and makes no assumptions about the drawing primitives that have been implemented.

1. Introduction

One of the chief advantages of ray tracing is that it provides a mathematically simple method for rendering accurate refractions through dielectrics [Whitted 80]. However, most dielectric objects are part of a more complex scene and may be nested in other objects. For example, consider an ice cube partially submerged in a glass of water. The ice, water, and glass each have different indices of refraction that would change the ray direction differently. Moreover, the change in ray direction depends on both the refractive index of its current medium and of the medium it is passing into. If a ray was entering ice from water, it would change direction differently than if it was entering ice from air.

A common method for rendering nested dielectrics is to model the scene ensuring that no two objects overlap. This can be done either using constructive

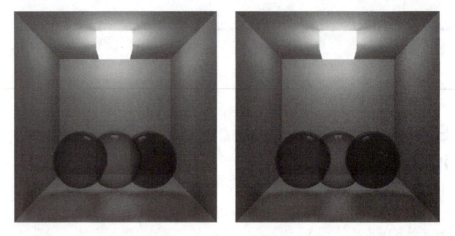

Figure 1. The same scene rendered with different priorities. In the first image, priorities decrease from left to right. In the second, the middle sphere has the lowest priority. To make the difference more visible, the spheres have been given the same refractive indices as the surrounding air. (See Color Plate XVI.)

solid geometry (CSG) or by manual manipulations of the geometry. A small gap is placed between objects to ensure that rays are never confused about which object they are hitting. This method presents a challenge in that the gap must be large enough so that floating point errors do not transpose the object borders, but if the gap is too large, it becomes a visible artifact in the rendered image. Our method allows nested dielectrics without requiring the renderer to support CSG primitives and without needing any gap between the nested objects. It can also allow certain pieces of geometry to be modeled at a lower resolution than would otherwise be necessary. Finally, the algorithm allows some surfaces to be ignored by the renderer, reducing the rendering time needed for some models.

2. Algorithm

Our method works by enforcing a strict hierarchy of closed geometry. All potentially overlapping materials are represented as closed solids and given a priority when they are defined. Our algorithm works by ensuring that if a ray is traveling through multiple objects, only the object with the highest priority will have any effect on the behavior of the ray. Essentially, the algorithm is a simplified form of CSG applied to the refraction problem in that object interfaces are defined by a geometric difference operation. This operation is controlled by the object priorities. Figure 1 demonstrates a scene using two different sets of priorities.

For our algorithm, nested objects should be modeled in such a way that ensures they overlap. For example, if rendering a glass filled with water, the boundary of the water would be set between the inner and outer walls of the glass. The modeler would assign a higher priority to the glass to ensure that, when a ray passed through the overlapping region, this region would be treated as part of the glass.

To determine which object a ray is effectively traveling through, the algorithm uses a simple structure called an interior list. Interior lists are small arrays stored with each ray that indicate which objects that ray is traveling through. Due to the fact that objects overlap, a ray's interior list may contain multiple objects. The highest priority object of a ray's interior list is the object which will influence the ray's behavior.

In order to handle the fact that objects overlap, all object intersections are evaluated using the interior list and priority numbers. Since only the highest priority object is considered to exist when multiple objects overlap, we have two cases: the ray intersects an object with a priority greater than or equal to the highest element in the ray's interior list (called a true intersection), or the ray intersects an object with a lower priority than this greatest interior list element (called a false intersection). Rays with empty interior lists will always produce true intersections. Examples of true and false intersections are shown in Figure 2.

This algorithm can be utilized in virtually any ray casting scheme including path tracing [Shirley 00] and photon mapping [Jensen 01], and should require only modest modifications to most existing renderers. These modifications are added to keep the interior list updated and to differentiate between true and false intersections.

2.1. False Ray Intersections

When a false intersection is encountered, no color calculations are performed and we simply continue searching for the next closest intersection ("color calculations" refer to the spawning of reflection and refraction rays, lighting, shadowing, and other similar calculations that would contribute to the color discovered by the given ray). This search is repeated until a true intersection is found or all possible intersections have been shown to be false, the latter indicating the ray missed all geometry.

The only computation made as a result of a false intersection is in the interior list. The intersected object is added to or removed from the ray's interior list based on whether the ray entered or exited this object, respectively.

2.2. True Ray Intersections

True intersections result in normal color calculations, just as they would in a normal ray tracer. Unlike a standard ray tracer, however, the reflection and

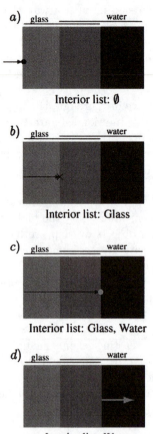

a) The ray intersects the glass from the outside. Since the ray did not begin in any object (the interior list is empty) this is guaranteed to be a true intersection. We would compute the color values for this point. The reflection ray would continue to use an empty interior list. The refraction ray is shown in b).

b) The refraction ray from a) continues into the glass. It next strikes the border of the water (entering the area where both water and glass are specified). Because the glass has a higher priority than water, the intersection with the water is a false intersection. The interior list is updated and the ray continues to search for an intersection.

c) The ray next strikes the other side of the glass. Because the glass is equal to the highest priority object in the interior list (itself) this is a true intersection. Color values for this point are calculated. The reflection ray's interior list would contain both the glass and the water objects. The refraction ray is shown in d).

d) The refraction ray from c) continues into the water.

Figure 2. True and false ray intersections. Glass (grey) has a higher priority than water (black). The dark grey area indicates where both materials overlap. Note that in a real image, the ray direction between a and b and between c and d would likely change due to refraction. (This was not done here to simplify the figure.) There would be no change in direction between b and c since the intersection in b is false.

refraction rays have interior lists that must be initialized. The reflection ray is simply given a copy of the original ray's interior list since the reflection ray crosses no additional boundaries. The refraction ray, however, is created by crossing from one object to another, and therefore would have a different interior list from the original. The refraction ray starts by copying the interior list of its parent, but then adds or removes the intersected object (depending on whether the refraction ray is entering or exiting this object, respectively).

At the same time, to compute the direction of the refracted ray it is necessary to know the refraction index of the current medium (the 'from-index') and of the medium the ray will be transitioning into (the 'to-index'). If the refraction ray is entering the intersected object, the from-index would be the index of the highest priority object in the original ray's interior list, and the to-index would be that of the intersected object. If the refraction ray is exiting the intersected object, the from-index would the index of the intersected object and the to-index would be the index of the highest priority object in the refraction ray's interior list. If a ray's interior list is empty, this indicates that ray is traveling outside all geometry. Usually this space is given the refractive index of air, although any index could be assigned.

3. Discussion

The key contribution of this algorithm is that it allows objects to overlap in model space while still producing correct borders in the rendered image. The method is relatively simple, but still manages to produce strong performance.

3.1. Advantages

This algorithm can significantly simplify the modeling of nested dielectrics. Consider a glass filled with water. Previously, the water would have been modeled as slightly smaller than the inside of the glass. In order to keep the gap between the objects as small as possible, the border of the water would need to be rendered at high resolution to closely follow the glass's surface. Using the method proposed in this paper, the sides of the water could be anywhere between the sides of the glass. Since the sides of the water would only be used to mark the water's boundary and would never be rendered, they could be modeled at a lower resolution. Only the glass boundaries would be modeled at high resolution because only these boundaries would be visible in the rendering.

A second advantage of this method is that it makes the modeling of some surfaces unnecessary. If a single surface forms the boundary between two objects, only the higher priority object needs to define this boundary. Consider gas bubbles completely surrounded by water. Previously it would have been necessary to model a border for the water surrounding the bubbles as well as modeling the bubbles themselves. However, using our technique, if one gives the gas bubbles a higher priority than the surrounding water only the bubbles would need to be modeled. As a result, careful ordering of priorities can actually reduce the number of boundaries against which intersection calculations must be performed.

A third advantage is that, because false intersections do not require color calculations, rendering time can actually be reduced in some models. Again, consider a glass with water in it. In a normal ray tracer, color values would be calculated when the ray entered the glass, exited the glass, and entered the water for a total of three calculations. Using our algorithm, one of these intersections would be false and would receive no further computation. As a result, the same set of ray intersections would result only in two color computations.

3.2. Implementation and Limitations

In order to keep track of which objects a ray is inside, our implementation simply toggles interior status of an object each time a ray crosses its boundary. If a ray intersects an object and the object is not in the interior list, then the ray must be entering the object. If the intersected object is already in the interior list, the ray is currently traveling through the object's interior and hence would exit the object at this point. This technique fails when the ray has a singular intersection with an object, such as along the object's silhouette. The use of a more sophisticated algorithm for determining which objects an ray was interior to at a given point could most likely eliminate these errors. However, even when using our simple toggle method, we found that standard anti-aliasing techniques removed most artifacts.

The primary disadvantage of the algorithm is the constraint that nested geometry must have overlapping borders. This requirement can provide some added complexity, especially if the surrounding material is very thin. In most cases, however, a simple scaling of the interior object by a small amount will be a sufficient solution.

3.3. Efficiency

Our algorithm requires very little additional overhead. In our implementation, when a false intersection is encountered, we simply move the base of the ray to the location of the false intersection and recast the ray in the same direction. This process is repeated until the nearest intersection is a true intersection. Despite the inefficiency of this method, we found that the use of this algorithm had a minimal impact on the overall time of the rendering and could, in some cases, actually reduce the rendering time. Specifically, in scenes with no dielectrics we found the algorithm to be only about 0.8% slower, while scenes with multiply-nested refractive objects, such as the glass in Figure 3, could have their overall rendering time reduced by more than 5% compared to an unmodified ray tracer.

Our implementation uses a short array of pointers to scene objects as the interior list. Because it is unlikely that a ray will start from inside more than

Figure 3. Glass with nested dielectrics. Priorities (from highest to lowest) are: the glass, the ice cube, the air bubbles under the water, and the water. (See Color Plate XV.)

a handful of objects, the interior list need only be large enough to contain a few elements. We found that the computations related to maintaining the interior list accounted for less than 0.4% of the runtime of the program, even for scenes with multiply nested objects.

Acknowledgments. Special thanks to Margarita Bratkova for creating the glass and ice cube models. This material is based upon work supported by the National Science Foundation under Grants: 9977218 and 9978099.

References

[Jensen 01] Henrik Wann Jensen. "Realistic Image Synthesis Using Photon Mapping," Natick, MA: A K Peters, Ltd., 2001.

[Shirley 00] Peter Shirley. "Realistic Ray Tracing," Natick, MA: A K Peters, Ltd., 2000.

[Whitted 80] Turner Whitted. "An Improved Illumination Model for Shaded Display," Communications of the ACM 23(6):343-349, 1980.

Web Information:

http://www.acm.org/jgt/papers/SchmidtBudge02

Charles Schmidt, University of Utah, School of Computing, 50 S. Central Campus Dr., Salt Lake City, UT, 84112 (cms@cs.utah.edu)

Brian Budge, University of Utah, School of Computing, 50 S. Central Campus Dr., Salt Lake City, UT, 84112 (budge@cs.utah.edu)

Received April 30, 2002; accepted in revised form May 15, 2002.

New Since Original Publication

A colleague referred us to the related work by Michael Gervautz [Gervautz 92]. This paper focuses on maintaining the correctness of material properties on boundaries of explicitly defined CSG objects using CSG trees. For example, the boundary shared by A and B is ambiguous; do you choose As material or Bs material to perform shading? The paper describes a method for doing these sorts of CSG operations consistently.

Dielectrics are mentioned in the paper, but only in the context that knowing which material to use at the interface could effect transmissions. Also, the idea of priorities is used, but they are implicitly defined by the order that the CSG object is defined ($A \cup B$ would use As material, whereas $B \cup A$ would use Bs), whereas our paper requires the user to explicitly define priorities.

Additional References

[Gervautz 92] Michael Gervautz. "Consistent Schemes for Addressing Surfaces when Ray Tracing Transparent CSG Objects," Computer Graphics Forum 11(4):203-211, 1992.

Current Contact Information:

Charles Schmidt, The MITRE Corp., 202 Burlington Rd., Bedford, MA 01730 (cms@cs.utah.edu)

Brian Budge, University of California, Davis, Institute for Data Analysis and Visualization, 2144 Academic Surge, Davis, CA 95616 (budge@cs.ucdavis.edu)

Part V

Rendering and Shadowing

Vol. 5, No. 1: 23–26

A Shaft Culling Tool

Eric Haines

Autodesk, Inc.

Abstract. Shaft culling is a means to accelerate the testing of visibility between two objects. This paper briefly describes an algorithm for shaft culling and various implementation options. The code and test harness for the algorithm is available online.

1. Introduction

Shaft culling is the process of examining two objects and determining what other objects are potentially in between them. The algorithm developed by Haines and Wallace [Haines, Wallace 94] was originally used to efficiently find the visibility between two objects for radiosity calculations. However, the method is a purely geometrical operation that can be used in other rendering techniques and for collision detection. The space swept out by a moving object (e.g., a catapult boulder) between one frame and the next forms a shaft, which can then be tested against the rest of the scene for collision. Another use is for quickly identifying all the objects between a point light source and some set of receivers. This can be helpful in more rapidly generating drop shadows by identifying only those objects that could project onto the receivers.

Putting an axis-aligned bounding box around both objects (called the shaft box) and connecting the two objects boxes by a set of planes forms a shaft (Figure 1). Other bounding volumes can then be tested against this shaft to see if they are fully outside. If not, additional testing can be done to see if they are fully inside. Testing a box against a shaft is a fast operation, as only one corner of the box needs to be compared to each particular plane in the shaft. For example, only the lower right corner c of the test box T in

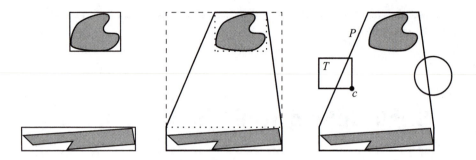

Figure 1. (a) Two object boxes will form the shaft; (b) shaft box and planes contain and connect boxes; (c) shaft/box and shaft/sphere testing.

Figure 1(c) needs to be tested against the shaft plane P to determine if the box is outside of the plane. The planes normal determines in advance which box corner to test. The signs of the normals coordinates determine the octant that corresponds to which box corner to test.

Code in C for this tool, along with a test harness that gathers statistics and shows the shafts formed, is available online at http://www.acm.org/jgt/. The original algorithm tested only boxes against the shaft; the code here includes support for testing spheres against the shaft. The sphere test is imperfect, but conservative: on occasion it will miscategorize a sphere as overlapping that is actually fully outside the shaft. For most applications of this tool this will lead to a little inefficiency, but not incorrect results.

2. Shaft Formation

The most involved part of shaft culling code is in forming the shaft itself. Efficiently generating the planes between the two object boxes is not an obvious process. The approach taken here can be thought of as follows. Each of the six faces of the shaft box is touched by one or both of the object boxes. Say we paint red each face touched by only the first object's box, and paint blue each face touched only by the second object's box. A shaft plane is formed for any edge whose two neighboring faces differ in color. The corresponding edges on the two object boxes are used to generate the plane. There will be from zero to eight shaft planes generated. As an example, in Figure 1(b) say the top object box is red and the bottom blue. The shaft's box top is then painted red and the other three faces are made blue. This leads to the formation of two shaft planes corresponding to the two upper corners, where the different colors meet.

```
For all faces f = (lo/hi.x/y/z) of the two object boxes
    /* color the face purple if box extents are the same */
    If the two box coordinates are equal:
       match[f] = true
    /* else color the face red or blue */
    Else:
       match[f] = false
       Note which object box coordinate is further out
       by saving color[f] = object box 1 or 2.


/* loop through and check all faces against each
 * other to determine if a plane should join them */
For face f1 = lo.x/y/z through hi.x/y (0 through 4)
    If match[f1] is false:
    /* else face is purple, and no edge forms a shaft */
       For face f2 = f1 + 1 through hi.z (through 5)
          If match[f2] is false:
          /* else face is purple */
             /* check if faces share an edge; opposite face
              * pairs such as lo.y and hi.y do not */
             If f1 + 3 is not equal to f2:
                /* test if faces are different colors */
                If color[f1] is not equal to color[f2]:
                   Create and store a plane joining the two
                   object box faces f1 & f2 along their common edge.
```

Figure 2. Shaft formation pseudocode.

The code gives two forms of this algorithm, one that finds these edges during run-time and another that uses a precomputed table which stores the relevant edges. The precomputed table has 64 entries, formed by six bits. Each bit represents the color of the corresponding shaft face, e.g., 0 for red, 1 for blue. Interestingly, on a Pentium II the precomputed table code forms a shaft in only 3% less time than the run-time code.

The run-time code has a further optimization. If both object boxes touch a shaft face, color that face purple. This corresponds to both object box faces lying in the same plane, meaning that no additional shaft planes are needed to join these faces. Again, only edges with one red and one blue neighbor will form a shaft; purple faces are ignored. This optimization avoids generating unnecessary shaft planes. It could be added to the table approach, but leads to a 729 entry table (3^6). Pseudocode for run-time shaft formation is given in Figure 2.

Once the shaft is formed, another optimization can be done. Each shaft plane chops off a certain volume of the shaft box. Logically, the shaft plane

Vol. 5, No. 1: 23–26

that cuts off the largest volume should be tested first, since it is most likely to cull out an arbitrary test box and so return more quickly. Code is included to sort the shafts planes by the amount of volume they cull. In practical benchmarks, this optimization was found to save an average of from 0.1 to 1.1 plane tests per non-trivial box/shaft comparison (i.e., ignoring cases where the test box was fully outside the shaft box). This savings is relatively minor, but if a single shaft is going to be tested against many boxes then this additional sorting step may pay off. More elaborate sorts could also be done; the code as given does not take into account the amount of overlap of volumes trimmed by the planes. For example, shaft planes for diagonally opposite edges could be good to pair up, as the trimmed volumes for these will never overlap.

A number of other variations on the code are outlined in the code comments, with tradeoffs in speed, memory, and flexibility discussed. Most of these code variations can be switched on or off and so can be tried by the user of this tool.

Acknowledgments. Thanks go to Martin Blais for checking the shaft code for errors and inefficiencies.

References

[Haines, Wallace 94] Eric A. Haines and John R. Wallace. "Shaft Culling for Efficient Ray-Traced Radiosity." *Photorealistic Rendering in Computer Graphics (Proceedings of the Second Eurographics Workshop on Rendering)*, pp. 122–138, New York: Springer-Verlag, 1994. Also in SIGGRAPH '91 *Frontiers in Rendering* course notes. Available online at http://www.acm.org/tog/editors/erich/

Web Information:

http://www.acm.org/jgt/papers/Haines00

Eric Haines, Autodesk, Inc., 1050 Craft Road, Ithaca, NY 14850 (erich@acm.org)

Received February 7, 2000; accepted March 15, 2000.

New Since Original Publication

This article was a distillation of thoughts and code I had developed after writing "Shaft Culling for Efficient Ray-Traced Radiosity." During 1994–1999, I consulted on an implementation by another programmer and thought more about the algorithm. I realized that there was a clearer way of describing how shafts are formed, and thus a cleaner way of coding the whole method. There were also a number of minor enhancements that I realized could be added to the basic concept. I felt it worthwhile to summarize these new ideas and to present a straightforward implementation in this article. The *journal of graphics tools* was a good fit, as they encourage writing about tools and their website could store the related code.

Current Contact Information:

Eric Haines, 111 Eastwood Avenue, Ithaca, NY 14850 (erich@acm.org)

Vol. 6, No. 1: 19–27

Soft Planar Shadows Using Plateaus

Eric Haines
Autodesk, Inc.

Abstract. This paper presents an algorithm for rapidly creating and rendering soft shadows on a plane. It has the advantages of rapid generation on current hard-ware, as well as high quality and a realistic feel on a perceptual level. The algorithm renders the soft shadow to a texture. It begins with a hard drop shadow, then uses the object's silhouette edges to generate penumbrae in a single pass. Though the shadows generated are physically inaccurate, this technique yields penumbrae that are smooth and perceptually convincing.

1. Introduction

Projective shadow techniques are commonly used in real-time rendering. Blinn [Blinn 88] discusses how to generate and render shadows by projecting the shadow casting object's polygons onto a ground plane. A related, more general method is to generate a texture containing the shadow cast, which is then projected onto the objects receiving the shadow [McReynolds et al. 99], [Nguyen 99], [Segal et al. 92].

One drawback of these methods is that the shadows have a hard edge, i.e., there are no penumbrae. This is often perceived as being unrealistic and prone to misinterpretation. A number of techniques have been developed to synthesize penumbrae for projective shadows. One method is to blur the shadow texture generated as a post-process. Soler and Sillion [Soler, Sillion 98] improve upon this technique by using the shape of the light itself as the filtering kernel, thus giving a variety of penumbrae types. However, such blurring is uniform over the shadow edges, which is valid only when the edges

Vol. 6, No. 1: 19–27

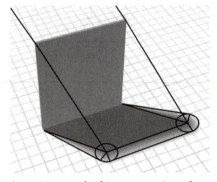

Figure 1. The object's vertices and edges are projected onto the ground. Silhouette vertices form radial gradient circles and edges form gradient quadrilaterals or triangles.

of the object casting the shadow are all the same distance from the shadow receiver. In reality, if an object touches the ground, the penumbra should be non-existent at this point, and widen out with the object's height (see Figure 1).

Heckbert and Herf [Heckbert, Herf 97], [Herf, Heckbert 96] take the approach of rendering the hard-edged planar projection of the shadow-caster a large number of times to a texture, sampling the area light source across its surface. The average of these shadows gives a smooth, realistic penumbra. This technique can also be done by using an accumulation buffer [Snyder et al. 92]. The main drawback of this method is that a large number of passes (typically 16 or more) are needed for smooth penumbrae; too few, and the hard-edged shadow edges do not blend, creating distracting artifacts. Gooch et al. [Gooch et al. 99] ameliorate these artifacts by generating a series of concentric shadows. However, multiple passes are still needed, shadow banding occurs, and visibly incorrect penumbrae are generated where the object touches the ground.

2. Algorithm

The algorithm presented here gives penumbrae in a single pass that are usually perceptually acceptable. The idea is to cast a hard drop shadow onto a texture, then create an approximation of the penumbra along the silhouette edges. The resulting image is used as a texture on the ground plane. This algorithm was inspired by the work of Parker et al. [Parker et al. 98], in which the authors use ray tracing for shadow testing and extend each shadowing object's edges outwards as gradients, then use the darkest shadowing object found.

Figure 1 shows the basic idea behind the algorithm. The vertices and edges of the object are cast onto the ground plane. The interior of the projected object is rendered in the darkest color, as is normally done when casting a hard shadow. Where a silhouette vertex or edge lands is where the soft shadow is painted onto the ground plane. A vertex causes a radial gradient circle to be painted, with the vertex location being the darkest shade and going to white at the circle's edge. The radius of the circle is dependent on the height of the casting vertex above the ground plane. Similarly, each silhouette edge creates a quadrilateral or triangle, which is shaded from the darkest shade along the hard shadow edge and fades to white along the outer edge.

If these circles and polygons were drawn without regard to each other the results would be unconvincing at best, with light and dark areas intermingling at random. To implement Parker's idea of rendering the darkest shadowing object, the two-dimensional circle and polygonal painting primitives are turned into three-dimensional cones and hyperbolic sheets, with the darkness of the object determining its closeness in the Z-buffer. In this way when two drawing primitives overlap, the darkest one will always then be the one seen.

The algorithm consists of drawing three-dimensional primitives to form a texture approximating the soft shadow. First the frame buffer where the shadow texture image will be formed is cleared to white, and the Z-buffer cleared as usual. The view for this shadow texture is set to look straight down onto the ground plane (i.e., a plan view).

Next, shadow casting objects in the scene are projected onto the ground plane to create the umbra. In addition to rendering the projected polygons to the color buffer in the darkest desired shadow color, these polygons are also rendered at a depth equal to the hither plane and their z-depths are stored in the Z-buffer. In this way they are the closest objects and so will cover any penumbra-forming objects, thereby ensuring that the darkest shadowing object is visible in the texture.

All silhouette edges of the shadow-casters compared to the light source are then found. There are a variety of strategies for doing this process [Sander et al. 00]. A simple scheme is to create a list of edges for each shadow-casting object, with each edge pointing to the two polygons it shares. Each edge is then compared to the light source and if one polygon faces the light and the other faces away, then the edge is a silhouette edge. The silhouette edges are not needed in any particular order, so the processing cost is minimal.

The heart of the algorithm is drawing the penumbrae. For each projected silhouette-edge vertex, draw a cone with the tip pointing directly toward the viewer. The apex of the cone is at the hither, at the vertex's projected location on the texture. The base of the cone is centered around this shadow-cast location, and at the yon plane. The radius of the cone is proportional to the height of the original vertex above the ground. The cone is shaded by using a color per vertex: the apex is black and the base is white.

Figure 2. To ensure that the darkest gradient color covers overlapping lighter colors, the gradient circles are turned into cones and the quadrilaterals into three-dimensional sheets. The upper left image shows a side view of the three-dimensional object formed to create the shadow texture. The hard drop shadow forms the plateau's top, with cones and sheets forming the penumbra sides. The upper right image shows the gradient coloration applied to the plateau object's primitives. In the lower left this plateau object is then rendered from above, creating a soft shadow texture. The lower right is the final result, made by applying the shadow texture to the ground plane.

For each edge, the two cones formed by the edge's vertices are connected by a sheet at their outer tangents. The sheet is a flat quadrilateral when the cones have the same radius (i.e., are at the same height above the ground). When the radii differ, the quadrilateral's tangent edges are not in the same plane and the sheet formed is a hyperboloid. Each sheet is rendered with its closest edge (on the hither) as black and its far edge (on the yon) as white. Figure 2 shows the objects formed and rendered.

The resulting image is a reasonable approximation of a penumbra for the object (see Figure 3). This grayscale texture can be used as a projected lightmap to modify the illumination on the receiving surfaces. Shadow receivers do not have to be planar, but the further removed they are from the target plane used to generate the shadow texture, the less convincing the effect.

Figure 3. The image on the left shows a hard shadow; the middle image shows the effect of a small area light source; the image on the right shows a larger light source.

3. Implementation

A few details are worth mentioning in controlling and implementing this algorithm on graphics hardware.

The radii of the cones generated depend on the height of the vertex above the ground plane times some constant factor. This factor is proportional to the relative size of the area light source. In most of the images in this paper a value of 0.1 was used, i.e., the penumbra's width is one tenth of the height of the object casting the shadow.

Plateau objects do not have to extend from the hither to yon. In reality, the number of bits needed for z-depths does not have to exceed the number of gray levels displayable, since the depths are only used to differentiate these levels.

Since each silhouette edge is processed separately, a shared vertex would normally generate its cone twice. A mailbox scheme can be used to render the cone for each silhouette vertex only once. When an edge generates two cones and a sheet, each vertex record used stores the current frame number. If another silhouette edge is encountered that would re-render one of these vertices, the vertex can be skipped if the current frame number is equal to its stored frame number.

Sheets are generated by connecting two cones at their tangents. These tangent edges are easy to find when the cones have the same radius: the first vertex is at the tip of the cone, and the second vertex is on the base's edge at a point on a line perpendicular to the silhouette edge. When the cone radii differ, the tangent line between the two base circles is computed and the points used to generate the tangent edges [Glassner 90]. See Figure 4.

If only two triangles are used to render a hyperboloid sheet, the penumbra can contain artifacts. It is worth tessellating such sheets when the radii differ by a noticeable factor. In Figure 2 the left side of the plateau (shown in the

211

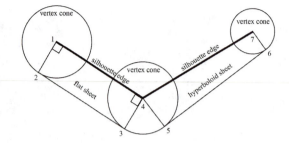

Figure 4. Sheet formation. If the cones have the same radius (i.e., the vertices are at the same height above the plane), the sheet 1–2–3–4 is formed by generating points along lines perpendicular to the original edge 1–4. If the radii differ, the sheet edges 4–5–6–7 are generated at the tangent points for line 5–6

upper images) corresponds with the penumbra formed by the left edge of the standing slab. This sheet is tessellated in order to avoid warping artifacts due to Gouraud interpolation [Woo et al. 96]. Such sheets are a type of Coons surface, bounded and defined by four straight lines. The surface is tessellated by linearly interpolating along each axis and generating the grid points.

A number of techniques can be used to shade gradient polygons. Using a color per vertex is one method. Another is to turn linear fog on, going from darkest gray for the objects at the hither to full white at the yon. A third is to use the Z-buffer depth values directly as grayscale color; only 8 bits of z-depth are needed in any case. A fourth method is to apply a one-dimensional grayscale texture to the plateau object. This last method allows other falloff patterns to be used in the penumbra. As Parker et al. [Parker et al. 98] note, the intensity gradient of the penumbra for a disk area light is sinusoidal, not linear. So, by using a texture map which translates the height of plateau surfaces into a sine curve falloff, a more realistic penumbra can be achieved, while also avoiding Mach banding at the penumbra's borders. See Figure 5.

Figure 5. In the image on the left, the penumbra is formed with a linear gradient. The middle image shows a sinusoidal drop off. The image on the right is a Heckbert and Herf style rendering for comparison.

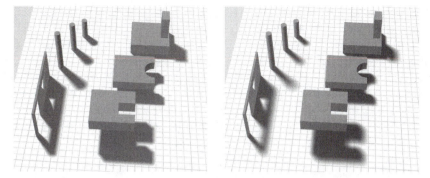

Figure 6. Plateau shadows on the left image, Heckbert and Herf shadows on the right. A Monte-Carlo sampled disk area light source using 256 shadow passes was used for Heckbert and Herf.

4. Discussion

The major drawback of the algorithm is that shadows always have fully dark umbra regions, similar to Parker et al. [Parker et al. 98]. Each object will cast a hard shadow umbra, no matter how far it is from the ground plane. Using a proper sinusoidal falloff has the effect of making the umbra appear even larger overall (see Figure 5). Another artifact is that where silhouette edges meet to form a concavity there is a sharp, unnatural change in the penumbra (see Figure 3).

Figure 6 shows comparable views of the same scene using plateaus and Heckbert and Herf's scheme. The image on the right is more physically correct and shows where the plateau scheme is a poor match. There are still some overlap artifacts visible when using the Heckbert and Herf algorithm, even with 256 passes (See the images at the web site listed at the end of this paper.)

Variants of the algorithm are possible. Instead of detecting silhouette edges, all polygons (or, somewhat more efficiently, all polygons facing the light) in a shadow caster could be used to generate cones and the two sheets joining the cones. While inefficient, this method has the advantage of needing no model connectivity information nor silhouette testing whatsoever.

Conversely, if full connectivity information is available, vertex cones could be trimmed in shape. By knowing the two silhouette edges sharing a vertex, some or all of the vertex cone's triangular faces are hidden and so do not need to be generated. For some (but not all) concave angles the cone is entirely hidden by the two edge sheets connected to it.

The basic algorithm makes the assumption that the light source is not near the horizon. As a light nears the horizon the penumbrae will become more non-uniform, i.e., they will become elongated. This stretching can be simulated by using and connecting cones with elliptical bases, with the major axis of the ellipse pointing towards the light.

Similar to Soler and Sillion's use of different filtering kernels to represent different light shapes [Soler, Sillion 98], different geometric projectors might be used. For example, instead of creating cones, pyramidal shapes with rounded corners might be used as a basis to represent rectangular area lights, or lozenge shapes might represent linear lights such as fluorescent tubes. No experiments have been done with such variants at this time, so it is unclear whether this is a viable approach.

References

[Blinn 88] Jim Blinn. "Me and My (Fake) Shadow." *IEEE Computer Graphics and Applications* 8(1): 82–86 (January 1988). Also collected in [Blinn 96].

[Blinn 96] Jim Blinn. *Jim Blinn's Corner: A Trip Down the Graphics Pipeline,* San Francisco: Morgan Kaufmann Publishers, Inc., 1996.

[Glassner 90] Andrew Glassner. "Useful 2D Geometry." In *Graphics Gems,* edited by Andrew S. Glassner, pp. 3–11, Cambridge, MA: Academic Press Inc., 1990.

[Gooch et al. 99] Bruce Gooch, Peter-Pike J. Sloan, Amy Gooch, Peter Shirley, and Richard Riesenfeld. "Interactive Technical Illustration." In *Proceedings 1999 ACM Symposium on Interactive 3D Graphics,* edited by Jessica K. Hodgins and James D. Foley, pp. 31–38, New York: ACM SIGGRAPH, April 1999. http://www.cs.utah.edu/~bgooch/ITI/

[Heckbert, Herf 97] Paul S. Heckbert and Michael Herf. *Simulating Soft Shadows with Graphics Hardware,* Technical Report CMU-CS-97-104, Carnegie Mellon University, January 1997.

[Herf, Heckbert 96] Michael Herf and Paul S. Heckbert. "Fast Soft Shadows." Sketch in *Visual Proceedings (SIGGRAPH '96),* p. 145, New York: ACM SIGGRAPH, August 1996.

[McReynolds et al. 99] Tom McReynolds, David Blythe, Brad Grantham, and Scott Nelson. "Advanced Graphics Programming Techniques Using OpenGL." *Course notes at SIGGRAPH '99,* New York: ACM SIGGRAPH, 1999. http://reality.sgi.com/blythe/sig99/

[Nguyen 99] Hubert Huu Nguyen. "Casting Shadows on Volumes." *Game Developer* 6(3): 44–53 (March 1999).

[Parker et al. 98] Steven Parker, Peter Shirley, and Brian Smits. *Single Sample Soft Shadows,* Tech. Rep. UUCS-98-019, Computer Science Department, University of Utah, October 1998.
http://www.cs.utah.edu/~bes/papers/coneShadow/

[Sander et al. 00] Pedro V. Sander, Xianfeng Gu, Steven J. Gortler, Hugues Hoppe, and John Snyder. "Silhouette Clipping." In *Proceedings of SIGGRAPH 2000, Computer Graphics Proceedings, Annual Conference Series,* edited by Kurt Akeley, pp. 327–334, Reading, MA: Addison-Wesley, 2000. `http://www.deas.harvard.edu/~pvs/research/silclip/`

[Segal et al. 92] M. Segal, C. Korobkin, R. van Widenfelt, J. Foran, and P. Haeberli. "Fast Shadows and Lighting Effects Using Texture Mapping." *Computer Graphics (Proc. SIGGRAPH '92)* 26(2): 249–252 (July 1992).

[Snyder et al. 92] John Snyder, Ronen Barzel, and Steve Gabriel. "Motion Blur on Graphics Workstations." *Graphics Gems III,* ed. David Kirk, Academic Press, pp. 374–382, 1992.

[Soler, Sillion 98] Cyril Soler and Francois Sillion. "Fast Calculation of Soft Shadow Textures Using Convolution." In *Proceedings of SIGGRAPH 98, Computer Graphics Proceedings, Annual Conference Series,* edited by Michael Cohen, pp. 321–332, Reading, MA: Addison Wesley, 1998. `http://www-images.imag.fr/Membres/Francois.Sillion/Papers/Index.html`

[Woo et al. 96] Andrew Woo, Andrew Pearce, and Marc Ouellette. "It's Really Not a Rendering Bug, You See... ." *IEEE Computer Graphics and Applications* 16(5): 21–25 (September 1996). `http://computer.org/cga/cg1996/g5toc.htm`

Web Information:

Sample images showing a comparison of the technique presented in this paper and those of Heckbert/Herf are available online at `http://www.acm.org/jgt/papers/Haines01`.

Eric Haines, Autodesk, Inc., 1050 Craft Road, Ithaca, New York 14850 (erich@acm.org)

Received March 13, 2001; accepted in revised form July 20, 2001.

New Since Original Publication

At the time I wrote this paper I considered it an interesting method, but one with limited applications, since it worked for only planar receivers. Happily, it helped inspire Tomas Akenine-Möller and Ulf Assarsson [Akenine-Möller, Assarsson 02] to consider the problem. Out of this research came their penumbra wedges algorithm, an extension of shadow volumes that can run on the graphics accelerator. In other related work, Wyman and Hansen [Wyman,

Hansen 03] created the idea of penumbra maps, a clever way of casting soft shadows on arbitrary surfaces by extending shadow maps with a buffer containing penumbra information. The moral of the story, for me, is that it is worth publishing even a relatively minor new way of thinking, as it "adds a brick to the palace," that is, it can help provide opportunities for others to create even better methods. I am appreciative that the *journal of graphics tools* has provided a venue for small articles such as mine.

Additional References

[Akenine-Möller, Assarsson 02] Tomas Akenine-Möller and Ulf Assarsson. "Approximate Soft Shadows on Arbitrary Surfaces Using Penumbra Wedges." In *EGRW 02: Proceedings of the 13th Eurographics Workshop on Rendering,* edited by Paul Debevec and Simon Gibson, pp. 297–306, Aire-la-Ville, Switzerland: Eurographics Association, 2002.

[Wyman, Hansen 03] Chris Wyman and Charles Hansen. "Penumbra Maps: Approximate Soft Shadows in Real-Time." In *EGRW 03: Proceedings of the 14th Eurographics Workshop on Rendering,* edited by Per Christensen and Daniel Cohen-Or, pp. 202–207, Aire-la-Ville, Switzerland: Eurographics Association, 2003.

Current Contact Information:

Eric Haines, 111 Eastwood Avenue, Ithaca, NY 14850 (erich@acm.org)

Vol. 7, No. 4: 9–18

Practical Shadow Mapping

Stefan Brabec, Thomas Annen, Hans-Peter Seidel

Max-Planck-Institut für Informatik

Abstract. In this paper, we present several methods that can greatly improve image quality when using the shadow mapping algorithm. Shadow artifacts introduced by shadow mapping are mainly due to low resolution shadow maps and/or the limited numerical precision used when performing the shadow test. These problems especially arise when the light source's viewing frustum, from which the shadow map is generated, is not adjusted to the actual camera view. We show how a tight-fitting frustum can be computed such that the shadow mapping algorithm concentrates on the visible parts of the scene and takes advantage of nearly the full available precision. Furthermore, we recommend uniformly spaced depth values in contrast to perspectively spaced depths in order to equally sample the scene seen from the light source.

1. Introduction

Shadow mapping [Williams 78], [Segal et al. 92] is one of the most common shadow techniques used for real time rendering. In recent years, dedicated hardware support for shadow mapping made its way from high-end visualization systems to consumer class graphics hardware. In order to obtain visually pleasing shadows, one has to be careful about sampling artifacts that are likely to occur. These artifacts especially arise when the light source's viewing frustum is not adapted to the current camera view. Recently, two papers addressed this issue. Fernando et al. [Fernando et al. 01] came up with a method called *adaptive shadow maps* (ASMs) where they presented a hierarchical refinement structure that adaptively generates shadow maps based on the camera view. ASMs can be used for hardware-accelerated rendering, but require many rendering passes in order to refine the shadow map. Stamminger et al. [Stamminger and Drettakis 02] showed that it is also possible to compute shadow maps in the post-perspective space of the current camera view. These *perspective shadow maps* (PSMs) can be directly implemented in hardware and greatly reduce shadow map aliasing. Drawbacks of the PSMs are that the shadow quality varies strongly with the setting of the camera's near and far plane and that special cases have to be handled if, e.g., the light source is located behind the viewer.

In this paper, we focus on the traditional shadow map algorithm and show how the light source's viewing frustum can be adjusted to use most of the available precision, in terms of shadow map resolution. Since it is also important that the available depth precision is used equally for all regions inside this frustum, we show how uniformly spaced depth values can be used when generating the shadow map.

Let us first recall how shadow mapping can be implemented using graphics hardware. In a first step, the scene is rendered as seen by the light source. Using the z-buffer, we obtain the depth values of the frontmost pixels which are then stored away in the so-called *shadow map*. In the second step, the scene is rendered once again, this time from the camera's point of view. To check whether a given pixel is in shadow, we transform the pixel's coordinates to the light source's coordinate system. By comparing the resulting distance value with the corresponding value stored in the shadow map, we can check if a pixel is in shadow or lit by the light source. This comparison step is illustrated

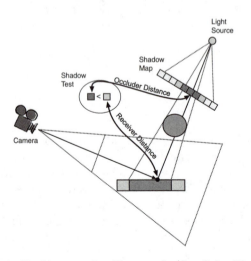

Figure 1. Shadow mapping illustrated. (See Color Plate VI.)

in Figure 1. Shadow mapping can be efficiently and directly implemented on graphics hardware [Segal et al. 92] with an internal texture format to store shadow maps and a comparison mechanism to perform the shadow test.

2. Distribution of Depth Values

When rendering the scene from a given viewpoint, depth values are sampled nonuniformly $(1/z)$ due to the perspective projection. This makes sense for the camera position, since objects near to the viewer are more important than those far away, and therefore sampled at a higher precision. For the light source position, this assumption is no longer true. It could be the case that objects very far from the light source are the main focus of the actual camera, so sampling those at lower z precision may introduce artifacts, e.g., missing shadow detail. A solution to this was presented by Heidrich [Heidrich 99]. Here, depth values are sampled uniformly using a 1D ramp texture that maps eye space depth values to color values, which are later used as the corresponding shadow map.

Figure 2 illustrates the difference between linear and $1/z$ mapping. On the left side, depth values are sampled using the traditional perspective projection. Objects near to the light source obtain most of the available depth values, whereas objects far away (e.g., the ground plane) have less precision. Shadow details for the teapot may be missing while the torus may be oversampled. The right side of Figure 2 shows the same setup using a linear distribution of depth values. Here, all objects are sampled equally. We can achieve this linear distribution of depth values using a customized vertex transformation, which can be implemented using the so-called *vertex shader* or *vertex program* functionality [Lindholm et al. 01] available on all recent graphics cards.

Instead of transforming all components of a homogeneous point $P = (x_e, y_e, z_e, w_e)$ by the perspective transformation matrix, e.g., $(x, y, z, w) = Light_{proj} \cdot P$, we replace the z component by a new value $z' = z_l * w$. The

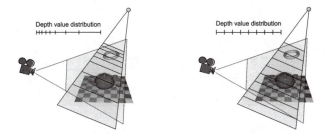

Figure 2. Distribution of depth values. Left: $1/z$. Right: z linear. (See Color Plate VII.)

linear depth value $z_l \in [0; 1]$ corresponds to the eye space value z_e mapped according to the light source near and far plane:

$$z_l = -\frac{z_e + near}{far - near}.$$

To account for normalization $(x/w, y/w, z/w, 1)$ which takes place afterwards (normalized device coordinates), we also premultiply z_l by w. This way, the z component is not affected by the perspective division, and depth values are uniformly distributed between the near and far plane.

3.   How Near, How Far?

Another very important property that affects the depth precision of the shadow map is the setting of the near and far plane when rendering from the light source position. A common approach is to set those to nearly arbitrary values like 0.01 and 1000.0 and hope that a shadow map with 24-bit precision will still cover enough of the relevant shadow information.

By analyzing the scene, we can improve depth precision by setting the near and far plane such that all relevant objects are inside the light's viewing frustum, as depicted on the left side of Figure 3.

In terms of depth precision, this setting is still far from being optimal. It is clear that the torus needs to be included in the shadow map, since it will cast a large shadow onto the ground plane and teapot, but for this shadow information, only one bit would be sufficient since the shadow caster itself is not seen by the camera. So what we really would like to have is some kind of tight-fitting near and far plane setup that concentrates on those objects that are visible in the final scene (seen from camera position). This optimal setting is depicted on the right side of Figure 3. If we would render this scene with the traditional approach, shadows cast by the torus would be missing since the whole object lies outside the light's viewing frustum and would be clipped away.

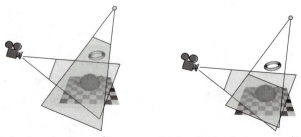

Figure 3. Left: Standard near/far setup. Right: Tight-fitting near/far setup. (See Color Plate VIII.)

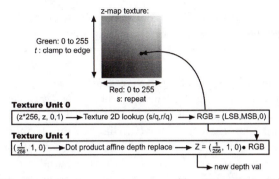

Figure 4. Texture shader setup. (See Color Plate V.)

We can easily include such objects by having depth values of objects in front of the near or beyond the far plane clamped to zero or one, respectively. This clamping can be achieved with a special *depth replace* texture mode, available as part of the *texture shader* extension provided by recent NVIDIA graphics cards [NVIDIA 02][1].

Assume we want to render a shadow map with 16-bit precision where depth values outside the valid range are clamped rather than clipped away. These depth values can be encoded using two bytes, where one contains the least significant bits (LSB) while the other stores the most significant bits (MSB). If we set up a two-dimensional ramp texture in which we encode the LSBs in the red channel (0 to 255, column) and in the green channel we store the MSBs (0 to 255, row position), we can map the lower 8 bit of a given z value by setting the s coordinate to $256.0 * z$ and using the s coordinate repeat mode. This way, s maps the fractional part of $256.0 * z$ to the LSB entry in the ramp texture. To code the MSB, we can directly map z to t and use a *clamp-to-edge* mode such that values $z < 0$ are clamped to 0 and values $z > 1.0$ are clamped to 1.0. To replace the current window space depth value with the new, clamped value, we now have to set up the texture shader as depicted in Figure 4. Texture unit 0 is responsible for mapping the higher and lower bits of the depth value to a color encoded RGB value (blue component set to zero). Texture unit 1 is configured to perform a *dot product depth replace* operation, which takes texture coordinates from unit 1 and computes a dot product with the result of the previous texture unit (color encoded depth). The result is a new depth value that is just a clamped version of the original depth value.

One problem with this texture shader setup is that the LSB is repeated even for objects in front or beyond the near/far planes, due to the s coordinate texture repeat. If we set the planes such that all pixels in front of the near

---

[1] As the name implies, this texture mode replaces a fragment's window space depth value.

clipping plane are mapped to a MSB of 0 and pixels beyond the far clipping plane to a MSB of 255, we do not have to worry about the LSB part of the depth value. So the effective range of depth values is between 0x0100 and 0xfeff[2].

Up to now, we did not handle the view frustum culling that takes place before the rasterization. If, for example, an object lies completely in front of the near clipping plane, all triangles would be culled away after the transformation step (clip coordinates). To avoid this, we simply modify the vertex shader described in Section 2. such that the z component of the output position is set to a value of $0.5 * w$. This way, all vertices are forced to lie between the valid $[0; 1]$ z range. The z values passed as texture coordinates for texture unit 0 are still the linear z_l. After the depth replace step, we then restore valid z coordinates used for depth testing.

4.   Concentrating on the Visible Part

In the previous section, we discussed how important the setting of near and far clipping is with respect to the depth resolution. For the shadow map resolution (width and height), the remaining four sides of the light's viewing frustum are crucial.

Consider a very large scene and a spotlight with a large cutoff angle. If we would just render the shadow map using the cutoff angle to determine the view frustum, we would receive very coarse shadow edges when the camera focuses on small portions of the scene. Hence, it is important that the viewing frustum of the light is optimized for the current camera view. This can be achieved by determining the visible pixels (as seen from the camera) and constructing a viewing frustum that includes all these relevant pixels.

In order to compute the visible pixels, we first render the scene from the camera position and use projective texturing to map a control texture onto the scene. This control texture is projected from the light source position and contains color-coded information about the row-column position, similar to the ramp texture used for the depth replace. In this step, we use the maximal light frustum (cutoff angle) in order to ensure that all illuminated parts of the scene are processed. Since pixels outside the light frustum are not relevant, we reserve one row-column entry, e.g., $(0, 0)$, for outside regions and use this as the texture's border color. By reading back the frame buffer to host memory, we can now analyze which regions in the shadow map are used. In the following subsections, we will discuss methods for finding a suitable light frustum based on this information.

---

[2]This method can be extended to 24-bit depths by using a special HILO texture format [NVIDIA 02], for which filtering takes place at 16 bits per component.

4.1. Axis-Aligned Bounding Rectangle

The easiest and fastest method is to compute the axis-aligned bounding rectangle that encloses all relevant pixels. This can be implemented by searching for the maximum and minimum row and column values, while leaving out the values used for the outside part (texture border). This bounding rectangle can now be used to focus the shadow map on the visible pixels in the scene. All we have to do is to perform a scale and bias on the x and y coordinates after the light's projection matrix to bring

$$[x_{min}; x_{max}] \times [y_{min}; y_{max}] \rightarrow [-1; 1] \times [-1; 1].$$

4.2. Optimal Bounding Rectangle

A better solution for adjusting the view of the light source is to compute the optimal bounding rectangle that encloses all visible pixels. This can be realized by using a method known as the *rotating calipers* algorithm [Toussaint 83], [Pirzadeh 99] which is capable of computing the minimum area enclosing rectangle in linear time. We start by computing the two-dimensional convex hull of all visible points using the *monotone chain* algorithm presented by [Andrew 97].

As stated by [Pirzadeh 99], the minimum area rectangle enclosing a convex polygon P has a side collinear with an edge of P. Using this property, a brute-force approach would be to construct an enclosing rectangle for each edge of P. This has a complexity of $O(n^2)$ since we have to find minima and maxima for each edge separately. The rotating calipers algorithm rotates two sets of parallel lines (calipers) *around* the polygon and incrementally updates the extreme values, thus requiring only linear time to find the optimal bounding rectangle. Figure 5 illustrates one step of this algorithm: The support lines are rotated (clockwise) until a line coincides with an edge of P. If the area of the new bounding rectangle is less than the stored minimum area rectangle, this bounding rectangle becomes the new minimum. This procedure is repeated until the accumulated rotation angle is greater than 90 degrees [3].

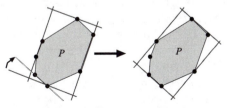

Figure 5. Rotating calipers.

[3]For a detailed description of the algorithm, please see [Pirzadeh 99].

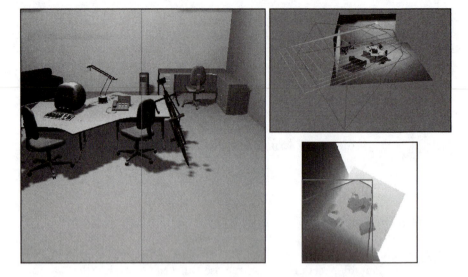

Figure 6. Left: Scene rendered with light frustum adjusted (left half) compared to the traditional method (right half). Top right: The optimized light frustum (green) and camera frustum (red). Bottom right: Location of convex hull (red) and minimum area enclosing rectangle (green) in the nonoptimized shadow map. (See Color Plate IX.)

5. Examples

Figure 6 shows an example scene illuminated by one spotlight with a large cutoff angle. Here, the image resolution was set to 512×512 pixels (4-times oversampling), whereas the shadow map only has a resolution of 256×256 pixels. For the control rendering pass a 64×64 pixel region was used. The frame rates for this scene are about 20 to 25 frames per second (74000 triangles). In the left image in Figure 6, we directly compared the adjusted light frustum (left half of the image) with the result obtained using some fixed setting (right half). Here, the adjustment can only slightly improve the shadow quality (coarse shadow edges), but the algorithm still computes an optimal setting for the light's near and far clipping plane. The top right image shows the optimized light frustum and the camera frustum, seen from a different perspective. In the bottom right image, the convex hull and the resulting minimum area enclosing rectangle are drawn as they are located in the nonoptimized shadow map.

In Figure 7, the camera was moved so that it focuses on a small part of the scene. Here, the automatic adjustment greatly improves shadow quality since the shadow map now also focuses on the important part of the scene. It can be seen that the light frustum is slightly overestimated, due to the resolution

Figure 7. Left: Camera close-up (left half with adjusted light frustum, right half without). Top right: Optimized light frustum (green) and camera frustum (red). Bottom right: nonoptimized shadow map and optimized shadow map. (See Color Plate X.)

of the control texture. An even tighter fit can be achieved by repeating the control texture rendering several times, so that the frustum converges near the optimum.

References

[Andrew 97] A.M. Andrew. "Another Efficient Algorithm for Convex Hulls in Two Dimensions." *Info. Proc. Letters* 9 (1997), 216–219.

[Fernando et al. 01] Randima Fernando, Sebastian Fernandez, Kavita Bala, and Donald P. Greenberg. "Adaptive Shadow Maps." In *Proceedings of ACM SIGGRAPH 2001*, pp. 387–390, August 2001.

[Heidrich 99] Wolfgang Heidrich. "High-Quality Shading and Lighting for Hardware-Accelerated Rendering." Ph.D. diss., University of Erlangen, 1999.

[Lindholm et al. 01] Erik Lindholm, Mark J. Kilgard, and Henry Moreton. "A User-Programmable Vertex Engine." In *Proceedings of SIGGRAPH 2001, Computer Graphics Proceedings, Annual Conference Series*, edited by E. Fiume, pp. 149–158, Reading, MA: Addison-Wesley, 2001.

[NVIDIA 02] NVIDIA Corporation. *NVIDIA OpenGL Extension Specifications*. Available from World Wide Web (http://www.nvidia.com), 2002.

[Pirzadeh 99] Hormoz Pirzadeh. "Computational Geometry with the Rotating Calipers." Master's thesis, McGill University, 1999.

[Stamminger and Drettakis 02] Marc Stamminger and George Drettakis. "Perspective Shadow Maps." *Proc. SIGGRAPH 2002, Transactions on Graphics* 21:3 (2002), 557–562.

[Segal et al. 92] Marc Segal, Carl Korobkin, Rolf van Widenfelt, Jim Foran, and Paul Haeberli. "Fast Shadow and Lighting Effects Using Texture Mapping." *Proc. SIGGRAPH '92, Computer Graphics* 26:2 (1992), 249–252.

[Toussaint 83] Godfried T. Toussaint. "Solving Geometric Problems with the Rotating Calipers." In *Proceedings of IEEE MELECON '83*, pp. A10.02/1–4. Los Alamitos, CA: IEEE Press, 1983.

[Williams 78] Lance Williams. "Casting Curved Shadows on Curved Surfaces." *Proc. SIGGRAPH '78, Computer Graphics* 12:3 (1978), 270–274.

Web Information:

http://www.acm.org/jgt/papers/BrabecAnnenSeidel02

Stefan Brabec, Max-Planck-Institut für Informatik, Stuhlsatzenhausweg 85, 66123 Saarbrücken, Germany (brabec@mpi-sb.mpg.de)

Thomas Annen, Max-Planck-Institut für Informatik, Stuhlsatzenhausweg 85, 66123 Saarbrücken, Germany (annen@mpi-sb.mpg.de)

Hans-Peter Seidel, Max-Planck-Institut für Informatik, Stuhlsatzenhausweg 85, 66123 Saarbrücken, Germany (hpseidel@mpi-sb.mpg.de)

Received April 2002; accepted September 2002.

New Since Original Publication

"Practical Shadow Mapping" introduced a new set of methods to improve the visual quality of applications based on the shadow mapping algorithm. The main goal of the paper was to describe novel and general concepts rather than specific hardware implementations. Although graphics hardware has become much faster and allows more flexible programming of vertex and fragment stages, most of the described concepts still apply.

However, using today's graphics chips the actual implementation is much easier, faster, and more general than before. The two-dimensional ramp texture is no longer needed, because fragment shaders allow direct output of depth values, making the complicated texture shader setup obsolete. Furthermore, floating point textures and frame buffers are now available on modern GPUs, which allow us to perform calculations with greater numerical precision. Even the computation of the optimal bounding rectangle should now be possible without any CPU-work, which would shift the algorithm from a hybrid GPU/CPU method to a fully hardware-accelerated one.

Current Contact Information:

Stefan Brabec, TomTec Imaging Systems GmbH, Edisonstr. 6, 85716 Unterschleissheim, Germany (sbrabec@tomtec.de)

Thomas Annen, Max-Planck-Institut für Informatik, Stuhlsatzenhausweg 85, 66123 Saarbrücken, Germany (tannen@mpi-sb.mpg.de)

Hans-Peter Seidel, Max-Planck-Institut für Informatik, Stuhlsatzenhausweg 85, 66123 Saarbrücken, Germany (hpseidel@mpi-sb.mpg.de)

Vol. 8, No. 3: 23–32

Optimized Shadow Mapping Using the Stencil Buffer

Jukka Arvo
Turku Centre for Computer Science and University of Turku

Timo Aila
Helsinki University of Technology and Hybrid Graphics, Ltd.

Abstract. Shadow maps and shadow volumes are common techniques for computing real-time shadows. We optimize the performance of a hardware-accelerated shadow mapping algorithm by rasterizing the light frustum into the stencil buffer, in a manner similar to the shadow volume algorithm. The pixel shader code that performs shadow tests and illumination computations is applied only to the pixels that are inside the light frustum. We also use deferred shading to further limit the operations to visible pixels. Our technique can be easily plugged into existing applications, and is especially useful for dynamic scenes that contain several local light sources. In our test scenarios, the overall frame rate was up to 2.2 times higher than for our comparison methods.

1. Introduction

A shadow map is a depth buffer created from the point of view of a light source [Williams 78]. With the standard shadow mapping algorithm, the shadow maps of all light sources must be accessed when a pixel is being shaded. This causes a lot of redundant work.

We present an acceleration technique that limits the shadow map tests to the visible pixels that are inside the frustum of the light source *(light frustum)*.

The technique works on currently available graphics hardware and uses a deferred shading [Deering et al. 88] approach, which first identifies the visible pixels, then accumulates the shadowed illumination, and finally computes the shading for each pixel. The high-level steps are:

- Compute a depth buffer from the viewpoint of the camera. At the same time, store the normal vectors at each pixel in a separate buffer. The normals are later used for shading computations.

- Create shadow maps for the light sources.

- For each light, use a technique similar to the shadow volumes [Crow 77] to create a stencil indicating which pixels are inside the light frustum. For those pixels only, perform the shadow map look-ups and compute the contribution of the light. Accumulate the diffuse and specular colors at each pixel in separate buffers.

- Do a final rendering pass that computes the shading of each visible pixel using the accumulated lighting contributions.

These steps are discussed in detail in Section 3. In our terminology, a *surface shader* is a pixel shader program that computes the final color of a pixel.

2.   Related Work

A point is in the umbra region, i.e., fully in shadow, if the line of sight from the point to the corresponding light source is blocked. Shadow volume algorithms construct quads that bound the umbra regions [Crow 77]. A point is in shadow if it is inside a geometrically bounded umbra region. Most hardware-based implementations of shadow volumes require a stencil buffer [Heidmann 91]. The stencil buffer is used for constructing a mask that indicates which pixels are inside the shadow volume. In the first pass, a depth buffer is generated from the viewpoint of the camera. The shadow volume quads are then rendered into the stencil buffer so that the visible pixels of the front-facing and back-facing shadow volume quads increment and decrement the stencil buffer values, respectively. Finally, the lighting is accumulated into the pixels where the stencil buffer value is zero.

The shadow map method [Williams 78] is based on the z-buffer algorithm. First, a depth buffer is generated from the point of view of the light source. This image, known as a *shadow map*, is a discretized representation of the scene geometry as seen by the light source. A pixel is in shadow if its depth value, transformed into the light-space, is greater than the corresponding shadow map depth value. Multiple shadow maps are required for omnidirectional light sources.

Reeves et al. [Reeves et al. 87] describe *percentage closer filtering*, which creates soft shadow boundaries that look plausible but are not physically based.

3. Optimizations Using the Stencil Buffer

In a brute-force approach, every rasterized pixel has to be tested against all shadow maps, which usually causes redundant work (Figure 1). However, this technique can be accelerated by using the stencil buffer. Ideally, the shadow map look-ups and lighting computations should be performed only for the pixels that are both visible from the viewpoint and inside the convex

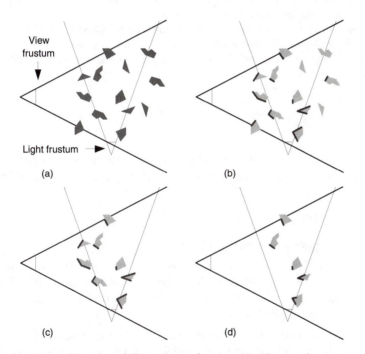

Figure 1. A two-dimensional illustration of regions that have to be processed when shadow maps are used. (a) A view frustum, a light frustum, and several objects. (b) The regions that are visible from the viewpoint are marked with bold red. (c) The visible regions that belong to objects that intersect with the light frustum. (d) The visible regions that are inside the light frustum and could thus be affected by the light source. Our algorithm processes only the regions marked in (d). (See Color Plate XI.)

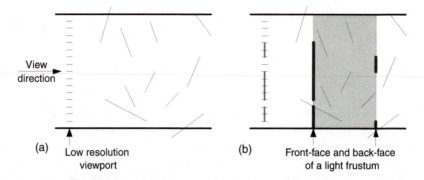

Figure 2. (a) This two-dimensional illustration shows a low-resolution viewport and several opaque objects (lines) in post-perspective space. The dark lines indicate the view frustum. (b) The visible parts of the front-face and the back-face of a light frustum are marked with bold red. The light source can affect only the parts of the objects that are inside the gray light frustum. In the visible subset of these parts, the front-face of the light frustum is visible (bold red) and the corresponding back-face is hidden. In practice, the test can be implemented using a stencil buffer. (See Color Plate XII.)

light frustum. The following explanation of our algorithm assumes per-pixel lighting:

1. Compute a depth buffer for the camera, clear the stencil buffer, and store the quantized normal vectors of the visible pixels in an additional buffer.

2. For each light source:

   (a) Compute a shadow map for the light source.

   (b) From the camera, rasterize the back-facing quads that bound the light frustum, using the depth buffer generated in Step 1. For the visible pixels, set the stencil buffer value to 1 (the right-most red lines in Figure 2(b)).

   (c) In the same fashion, rasterize the front-facing quads that bound the light frustum (the left-most red lines in Figure 2(b)). For the visible pixels, read the stencil buffer value. If the value is 1, both the front- and back-facing quads are visible and thus the stored depth value is behind the light frustum. In this case, clear the stencil buffer value. This ensures that the stencil buffer is left empty after all the quads have been rasterized. If the stencil value is 0, the pixel is inside the light frustum. In this case, execute a pixel shader program that performs the shadow map look-ups and per-pixel lighting using the screen-space (x, y)-coordinates, the depth buffer value, and the stored normal vector. Accumulate the lighting contributions in the diffuse and specular color buffers.

3. Render the geometry using materials so that the shadowed illumination is read from the diffuse and specular color buffers, and ambient lighting is accumulated.

The stencil test in Step 2(c) is performed before the pixel shader units, thus enabling the algorithm to early-exit the processing of redundant pixels. The amount of pixel processing can be further reduced for certain common light source types by constructing the light frustum according to the actual shape of the light source, using more than six planes.

3.1. Eye in Shadow

Like shadow volumes, our algorithm requires special treatment when the light frustum intersects with the near or far plane of the view frustum. Unlike shadow volumes, the light frustums are convex and can therefore be easily clipped before the rendering phase (Figure 3). The scene geometry should be rendered using a slightly reduced depth range in order to guarantee the correct rasterization of the parts of the light frustum that lie exactly on the near plane [Everitt and Kilgard 02]. For example, DirectX 9.0 offers *viewport scaling*, and OpenGL *glDepthRange()* for this purpose.

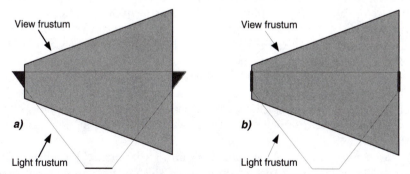

Figure 3. (a) A two-dimensional illustration of the light frustum intersecting with the near and far clipping planes of the view frustum. (b) The light frustum has been clipped by the near and far planes of the view frustum. The light frustum must remain closed after clipping, and thus the parts outside the near and far planes (marked with red) need to be projected to the planes by using parallel projection. (See Color Plate XIV.)

3.2. Usability and Filtering

Current graphics hardware can execute only a single pixel shader program per rendering pass, thereby forcing the replication of the shadowed illumination code into every lighting-dependent surface shader. With our method, the illumination computations are isolated into a separate pass that computes and stores the lighting coefficients for the subsequent material-rendering pass. This also has a minor performance advantage since the surface shaders are invariant of the number of affecting lights, and hence have to be switched less frequently.

The appearance of shadows created by using shadow maps can be improved by employing appropriate filtering methods, such as percentage closer filtering. Higher quality filtering increases the computational requirements. Thus, it makes sense to use lower quality filtering for the smaller or more distant parts of the scene. Having separated the filtering code from the surface shaders, we can adjust the filtering quality of shadows dynamically, according to the size and location of the light frustum, without modifying the surface shaders.

4. Results and Discussion

We have measured the average number of processed pixels and the resulting frame rates in three scenes (Figure 4) using a brute-force shadow mapping algorithm, an optimized variant, and our new stencil-based method. The brute-force algorithm computes shadowed illumination for every rasterized pixel during the processing of each light source. The optimized variant of the brute-force algorithm incorporates two optimizations. First, a depth buffer is computed for the viewpoint. This limits the subsequent illumination computations to the visible pixels. Then, for each light source, the objects outside the light frustum are culled and the shadowed illumination is computed only for the objects that intersect with the light frustum. This two-step optimization implements deferred shading, as the illumination passes only process visible pixels. However, a significant number of pixels outside the light frustum may still be processed because spatially large objects, such as a ground plane, often intersect many light frustums even if only a fraction of the model is affected by the light source.

All tests were run on a 2.4 GHz Pentium 4 with an ATI Radeon 9700 Pro graphics card. Both the screen and the shadow map resolutions were 1024×1024. All shadow maps were updated every frame in order to simulate a highly dynamic environment. Per-pixel lighting was computed according to the Phong model [Phong 75].

The results are shown in Tables 1 and 2. As expected, the brute-force algorithm was always the slowest. On average, our stencil-based algorithm processed roughly five times fewer pixels than the brute-force algorithm, and

Figure 4. The flashlight scene consists of 77K triangles and one spotlight. The room scene uses the same geometry, illuminated by one omnidirectional light source, implemented using six spotlights without the angular attenuation ramp. The parking lot scene has 101K triangles and 12 spotlights. (See Color Plate XIII.)

Scene	(tris/lights)	Number of processed pixels			Ratio in pixel processing	
		brute	optimized	stencil	brute / stencil	optimized / stencil
Flashlight	(77K/1)	1.03M	0.87M	0.38M	**2.71**	**2.29**
Room	(77K/6)	6.86M	5.74M	1.01M	**6.79**	**5.68**
Parking lot	(101K/12)	13.37M	8.42M	2.04M	**6.55**	**4.13**
PL split	(101K/12)	13.37M	5.29M	2.04M	**6.55**	**2.59**

Table 1. The average number of processed pixels per frame in three test scenes (Figure 4) for a brute-force (*brute*), an optimized variant (*optimized*), and for our new stencil-based improvement (*stencil*). Two versions of the parking lot scene were used. In the first one, the ground plane was a single object, and in the other version the plane was split into eight subobjects. The subdivision affects only the results of the optimized approach.

Scene	FPS: no filtering (2x2 percentage closer filter)			Ratio in FPS	
	brute	optimized	stencil	brute / stencil	optimized / stencil
Flashlight	32.70 (25.25)	36.42 (30.24)	39.72 (37.46)	**1.21 (1.48)**	**1.09 (1.24)**
Room	6.61 (5.27)	8.53 (6.69)	19.10 (15.58)	**2.88 (2.96)**	**2.24 (2.33)**
Parking lot	2.49 (2.11)	5.57 (4.38)	9.88 (8.80)	**4.01 (4.17)**	**1.77 (2.01)**
PL split	2.49 (2.11)	6.68 (5.36)	9.98 (8.80)	**4.01 (4.17)**	**1.49 (1.64)**

Table 2. The average frame rate without filtering, and with four-sample percentage closer filter in three test scenes (Figure 4) for a brute-force (*brute*), an optimized variant (*optimized*), and for our new stencil-based improvement (*stencil*). The second version of the parking lot scene had the ground plane split into eight subobjects. In all algorithms, the local illumination code consumed roughly half of the instructions in the pixel shading programs.

three times fewer than the optimized variant. The ratio in FPS between the stencil-based and the brute-force ranged from a modest 21% increase in the simplest flashlight scene to a factor of 4.0 in the parking lot scene. The FPS of the stencil-based method was 2.2 times higher in the room scene than with the optimized variant. The large performance difference with the room scene is due to the fact that many objects (walls, floor, ceiling) intersect with several light frustums.

A four-sample percentage closer filtering increased the performance difference by an additional 15%. The instruction count limitations in current graphics cards prevented using more samples, but that would of course have further amplified the difference. When the ground plane of the parking lot was not split into multiple subobjects, the performance difference increased by an additional 20%. Splitting all geometry into spatially compact pieces can be inconvenient in dynamic environments. Our method eliminates the need for such splitting. The algorithm does not pose limitations on the input geometry, and has a straightforward implementation on currently available graphics hardware.

Two factors reduced the overall performance of our method. Firstly, the additional diffuse, specular, and normal vector buffers consumed a total of $3 \times 1024 \times 1024 \times 4 \text{bytes} = 12$ MB of memory. The video memory bandwidth usage caused by the three additional buffers and the stencil buffer manipulation slowed our method down slightly. Secondly, stenciled shadow maps use computationally more involved pixel shader programs, and are therefore more dependent on the amount of computational resources in the pixel shader units.

Due to the transformation from image-space into light-space, our method is rather sensitive to the potentially limited accuracy of pixel shader implementations. In practice, floating-point arithmetic with at least 16 and preferably 24 bits of mantissa is required in order to avoid introducing additional noise.

As is often the case with z-buffer-based techniques, transparent surfaces require special treatment. In general, applications render the scenes by first rendering all the opaque objects and then all the sorted transparent surfaces with the z-buffer updates disabled. Our technique is applicable to rendering the opaque objects, whereas the transparent objects should be processed using other techniques.

Most shadow mapping implementations require additional checks for confirming that the processed pixel is not behind the light source, otherwise back-projection artifacts may occur. Such unwanted "phantom" lighting is automatically avoided in our method.

Acknowledgments. The authors would like to thank Mikko Lahti for the test scenes and Janne Kontkanen, Samuli Laine, Juha Lainema, Jaakko Lehtinen, Ville Miettinen, Lauri Savioja, and Jan Westerholm for helpful comments.

References

[Crow 77] Franklin Crow. "Shadow Algorithms for Computer Graphics." *Proc. SIGGRAPH '77, Computer Graphics* 11:2 (1977), 242–248.

[Deering et al. 88] Michael Deering, Stephanie Winner, Bic Schediwy, Chris Duffy, and Neil Hunt. "The Triangle Processor and Normal Vector Shader: A VLSI System for High Performance Graphics." *Proc. SIGGRAPH '88, Computer Graphics* 22:4 (1988), 21–30.

[Everitt and Kilgard 02] Cass Everitt and Mark Kilgard. "Practical and Robust Stenciled Shadow Volumes for Hardware-Accelerated Rendering." Available from World Wide Web (http://www.developer.nvidia.com), 2002.

[Heidmann 91] Tim Heidmann. "Real Shadows, Real Time." *Iris Universe* 18 (1991), 28–31.

[Phong 75] Bui-Tuong Phong. "Illumination for Computer Generated Pictures." *Communications of the ACM* 18:6 (1975), 311–317.

[Reeves et al. 87] William T. Reeves, David H. Salesin, and Robert L. Cook. "Rendering Antialiased Shadows with Depth Maps." *Proc. SIGGRAPH '87, Computer Graphics* 21:4 (1987), 283–291.

[Williams 78] Lance Williams. "Casting Curved Shadows on Curved Surfaces." *Proc. SIGGRAPH '78, Computer Graphics* 12:3 (1978), 270–274.

Web Information:

http://www.acm.org/jgt/papers/ArvoAila03

Jukka Arvo, Turku Centre for Computer Science, Lemminkäisenkatu 14A, 5th Floor, 20520 Turku, Finland (jukka.arvo@cs.utu.fi)

Timo Aila, Helsinki University of Technology, P.O. Box 5400, 02015 HUT, Finland (timo@hybrid.fi)

Received December 13, 2002; accepted in revised form July 21, 2003.

New Since Original Publication

A few interesting possibilities and one cube map issue were recognized after the original paper was published.

Normal culling. The number of processed pixels inside the light frustum can be reduced by utilizing normal culling, i.e., a technique that avoids processing the pixels whose normal vectors point away from the light source. This would be particularly useful for large percentage closer filtering kernels, because many shadow map lookups could be avoided in pixels that lie in umbra according to the local illumination model. Perhaps the easiest way to implement normal culling would be to test the orientations of the stored normal vectors when rasterizing the back faces of the light frustum (Step 2b of the algorithm) and to prevent setting stencil buffer values for pixels that represent surfaces facing away from the light source.

Clip planes. Deferred shading and multisampling do not work well together. Applications that do not use deferred shading could alternatively bound the light frustum using additional clip planes instead of an image-space computation mask. However, if the underlying graphics hardware really clips the triangles using the additional clip planes (instead of performing the clipping in the pixel pipeline), accuracy issues may arise. It might be possible to solve these problems by using a bias of some kind in the depth test.

Hierarchical culling. Current graphics hardware does not support hierarchical stencil buffers, but variants of the hierarchical z-buffer are widely implemented. Therefore, it might be more efficient to store the computation mask to the depth buffer instead. However, performance improvements could be achieved only if the depth mask generation does not take longer than the time gained by hierarchical pixel culling.

Omni-directional light sources and cube maps. In our test scenes we used six shadow maps to model an omni-directional light source and performed the lookups separately for each shadow map. Alternatively, the shadow maps could be represented as a cube map. This would reduce the number of redundant shadow map lookups in the comparison method and improve its performance. However, this would affect the results in only one of the tested scenes. Our method could be adapted to support cube maps by replacing the light frustum with a bounding volume that encloses the regions affected by the light source.

Current Contact Information:

Jukka Arvo, Turku Centre for Computer Science, Lemminkäisenkatu 14A, 5th Floor, 20520 Turku, Finland (jukka.arvo@cs.utu.fi)

Timo Aila, Helsinki University of Technology, P.O. Box 5400, 02015 HUT, Finland (timo@hybrid.fi)

Vol. 4, No. 3: 1–10

Faster Photon Map Global Illumination

Per H. Christensen

Abstract. The photon map method is an extension of ray tracing that makes it able to efficiently compute caustics and soft indirect illumination on surfaces and in participating media. This paper describes a method to further speed up the computation of soft indirect illumination (diffuse-diffuse light transport such as color bleeding) on surfaces. The speed-up is based on the observation that the many look-ups in the global photon map during final gathering can be simplified by precomputing local irradiance values at the photon positions. Our tests indicate that the calculation of soft indirect illumination during rendering, which is the most time-consuming part, can be sped up by a factor of 5–7 in typical scenes at the expense of 1) a precomputation that takes about 2%–5% of the time saved during rendering and 2) a 28% increase of memory use.

1. Background: The Photon Map Method

The photon map method [Jensen 95, Jensen 96a, Jensen 96b, Jensen 96c] has the following desirable properties for computation of global illumination:

1. It can handle all combinations of specular, glossy, and diffuse reflection and transmission (including caustics).

2. Since the photon map is independent of surface representation, the method can handle very complex scenes, instanced geometry, and implicit and procedural geometry.

3. It is relatively fast.

4. It is simple to parallelize.

The photon map method consists of three steps: photon tracing, photon map sorting, and rendering.

1.1. Photon Tracing

In the photon tracing step, photons are emitted from the light sources and traced through the scene using Monte Carlo simulation and Russian roulette [Glassner 95]. When a photon hits an object, it can either be reflected, transmitted, or absorbed (decided probabilistically based on the material parameters of the surface).

During the photon tracing, the photons are stored at the diffuse surfaces they intersect on their path. Note that each emitted photon can be stored several times along its path. The information stored is the position, incident direction, power of the photon, and some flags. (Jensen showed how this information can be stored in 20 bytes [Jensen 96b], but it turns out that 18 bytes are actually sufficient here: 12 bytes for position, 4 bytes for compressed power [Ward 91], 1 byte to encode incident direction, and 1 byte for the flags.) This collection of information about the photon paths is the *global photon map*.

In addition, the photons that hit a diffuse surface coming from a specular surface are stored in *a caustic photon map*.

1.2. Sorting the Photon Maps

In the second step, the photons stored in the two photon maps are sorted so that it is quick to locate the photons that are nearest a given point. A kd-tree [Bentley 75, de Berg et al. 97] is used since it is a compact representation that allows fast look-up and gracefully handles highly nonuniform distributions. This step is usually by far the fastest of the three.

1.3. Rendering

In the rendering step, specular reflections are computed with standard ray tracing [Whitted 80, Glassner 89]. However, when a ray intersects a diffuse surface, the incident illumination is computed in one of two ways [Jensen 96a]:

1. If the importance (that is, the contribution to the rendered image [Smits et al. 92, Christensen et al. 96]) is low, the total irradiance at the point is estimated from the power and density of the photons nearest to the point. (The procedure can be imagined as expanding a sphere

around the point until it contains n photons and then using these n photons to estimate the irradiance.) The global photon map is used for this. Typically, 30–200 photons are used.

2. If the importance is high, direct illumination is computed by sampling the light sources, caustics (indirect illumination from specular surfaces) are computed from the caustic photon map, and soft indirect illumination (indirect illumination from diffuse surfaces) is computed with final gathering [Reichert 92, Christensen et al. 96] (a single level of distribution ray tracing [Cook et al. 84]) using the global photon map.

In a final gather, the hemisphere above the point is sampled by shooting many rays and computing the radiance at the intersection points. For efficiency, the hemisphere can be importance-sampled based on the incident direction of nearby (indirect) photons and the BRDF of the surface [Jensen 95], or it can be adaptively sampled [Ward, Shakespeare 98], or both. Either way, most final gather rays are shot in the directions that are most likely to contribute most to the illumination. The radiances at the points that the final gather rays hit are usually estimated using the global photon map (using the same method as described under 1. above), but with one exception: If the distance along the final gather ray is below a given threshold, a new simpler final gathering is done [Jensen 96a]. Very few final gather rays are needed in these secondary final gathers—we have found 12 to be sufficient.

Each final gather is quite time-consuming since many hundred rays are shot, but fortunately relatively few final gathers are needed since previous results can be stored and used for interpolation and extrapolation of the soft indirect illumination [Ward, Heckbert 92].

2.   Faster Final Gathering

The most time-consuming part of rendering is the final gathering, in particular the computation of irradiance at the locations where final gather rays hit. During final gathering, the irradiance is computed repeatedly at nearly the same places—with rays from different final gathers hitting near each other. So, if we precompute the irradiance at selected locations, a lot of time can be saved.

2.1.   Precomputation of Irradiances

One could precompute the irradiances at fixed distances on each surface. But that would adapt poorly to the large differences in irradiance variation from place to place. Instead, the photon positions are an ideal choice for precomputing irradiances since the photons are inherently dense in areas with large illumination variation. The computed irradiance and surface normal at the photon position can be stored along with the other photon information in the photon map; this increases storage with 5 bytes per photon (4 bytes for compressed irradiance [Ward 91] and 1 byte for encoding the direction of the surface normal). Using 23 bytes per photon instead of 18 is a storage increase of less than 28%.

Estimating the irradiance at a point takes time $O(\log N)$, where N is the number of photons in the global photon map. Precomputing irradiance at the N photon positions therefore takes $O(N \log N)$. To reduce the time spent on the precomputation of irradiances, one can choose to only compute the irradiance at some photon positions—for example every fourth photon. This reduces the precomputation time to one-fourth, but does not change the rendering time.

2.2. *Final Gathering using Precomputed Irradiances*

When the irradiance at a point is needed during final gathering, the nearest photon with a similar normal is found in the photon map and its precomputed irradiance is used. ("Similar" means that the dot product of the normals has to be larger than a certain threshold, for example 0.9.)

Finding the nearest photon with a similar normal is only slightly slower, on average, than finding the nearest photon irrespective of normal. In the typical case, in which a flat or gently curving surface is not immediately adjacent to another surface, the nearest photon will always have an appropriate normal. Then, the only extraneous computation is a single dot product to check that the two normals agree. This only takes a small fraction of the time that was used to search the kd-tree to find that photon. Close to edges, the nearest photon sometimes happens to be on the wrong surface (and have the wrong normal). In this case, the first photon found is rejected and the search continues. Usually, the second or third found photon will have an acceptable normal. The additional dot products and searching required for this relatively rare case do not influence overall running time significantly.

Using the precomputed irradiance for the nearest photon means that the irradiance used for final gathering is approximated as a piece-wise constant function. Formally, the photon positions divide the surfaces in the scene into a Voronoi diagram with a constant irradiance within each cell. This approximation is acceptable because the difference between the irradiance at neighboring photon positions is relatively small (since many photons are used to compute each irradiance) and because we only use it for final gathering above a certain distance.

The precomputed irradiance of the nearest photon is also used if the importance of the irradiance is sufficiently low.

2.3. Discussion

This method gives a speed-up as long as there are more final gather rays than selected photon positions. In a typical scene, there are many more final gather rays than photons—often a factor of 10 or even 100. Therefore, the total number of irradiance estimates can be reduced by the same factor. This translates into a speed-up in the final gathering by a factor of 5–7 in typical scenes at the expense of a precomputation that takes only a few percent of the time saved during rendering.

Precomputing irradiances can probably also give a significant speed-up for rendering of global illumination in participating media [Jensen, Christensen 98] if the number of selected photon positions is smaller than the number of ray marching steps.

3. Results

The following tests were done on a Sony Vaio laptop with a single 233-MHz Pentium processor and 32 megabytes of memory.

3.1. Cornell Box

As a first test scene, we used a "Cornell box" with a chrome and a glass sphere. Emitting 220,000 photons resulted in 400,000 photons stored in the global photon map and 35,000 photons in the caustic photon map; this photon tracing took 10 seconds. Sorting the stored photons into two kd-trees also took 10 seconds. The photons stored in the global photon map are visualized in Figure 1(a).

In Figure 1(b), irradiance was computed at each image sample point from local irradiance estimates using 50 photons. In Figure 1(c), the irradiance was precomputed at all 400,000 global photon positions (also using 50 photons) and the irradiance at the nearest photon was used. Precomputing the irradiances took 58 seconds. There is no visual difference between Figures 1(b) and 1(c). In Figure 1(d), the irradiance was precomputed at only 100,000 photon positions, which took 15 seconds. The differences between Figures 1(c) and 1(d) can only be seen upon close inspection, and Figure 1(d) is fully adequate for use in final gathering.

In Figure 1(e), the irradiances from Figure 1(d) are multiplied by the diffuse reflection coefficients at each image sample point to produce a radiance estimate. This is the scene as the final gathering "sees" it.

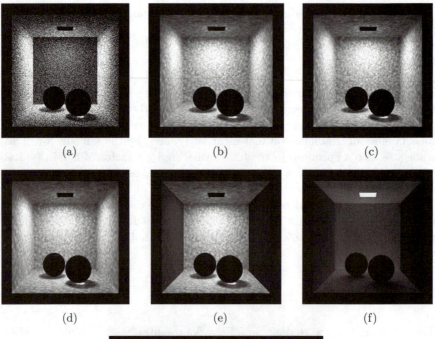

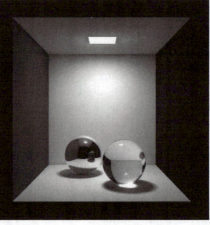

Figure 1. Cornell box with spheres: (a) Global photon map. (b) Irradiance estimates computed at image sample points (dimmed for display). (c) Precomputed irradiance estimates at all 400,000 photon positions (dimmed for display). (d) Precomputed irradiance estimates at 100,000 photon positions (dimmed for display). (e) Radiance estimates based on (d). (f) Soft indirect illumination computed with final gathering. (g) Complete image with direct illumination, specular reflection and refraction, caustics, and soft indirect illumination. (See Color Plate I.)

Figure 1(f) shows soft indirect illumination computed with final gathering based on radiance estimates as in Figure 1(e).

Figure 1(g) is the complete rendering. It contains direct illumination, specular reflection and refraction, caustics, and soft indirect illumination. Rendering this image at resolution 1024×1024 pixels with up to 16 samples per pixel took 5 minutes and 29 seconds. Out of this time, computing the soft indirect illumination took 2 minutes and 45 seconds. There were 11,800 final gathers with a total of 6.5 million final gather rays.

For comparison, rendering the same image without precomputing the irradiances takes approximately 20 minutes, out of which final gathering is responsible for 16 minutes. The resulting image is indistinguishable from Figure 1(g). In conclusion, precomputing the irradiances before rendering reduces the final gathering time by a factor of 5.8. The 15 seconds spent on precomputation constitute less than 2% of the 13 minutes and 15 seconds saved during final gathering.

Illumination effects to note in this image are the very smooth (purely indirect) illumination in the ceiling, the bright caustic under the glass sphere, and the three dim secondary caustics also under the glass sphere (focused light from the red, white, and blue bright spots on the walls).

3.2. *Interior Scene*

We also tested the method on a more complex scene, an interior scene with 1,050,000 polygons. Emitting and tracing 193,000 photons resulted in 500,000 photons stored in the global photon map. (We chose not to store photons in a caustic photon map for this scene since caustics are an orthogonal issue to the method described here.) Photon tracing took 39 seconds and sorting the photons took 12 seconds. The resulting photon map is shown in Figure 2(a).

In Figure 2(b), irradiance is computed at each image sample point from local irradiance estimates using 200 photons. In Figure 2(c), the irradiance was precomputed at all 500,000 photon positions (taking slightly less than 6 minutes) and the irradiance at the nearest photon was used. There is no visual difference between the images in Figure 2(b) and Figure 2(c).

In Figure 2(d), the irradiances from Figure 2(c) were multiplied by the diffuse reflection coefficient at each image sample point. This is the scene as the final gathering "sees" it.

Figure 2(e) is the complete rendering with direct illumination, specular reflection, and soft indirect illumination based on the precomputed irradiance estimates in Figure 2(c). Rendering this image at resolution 1024×768 pixels with up to 16 samples per pixel took 60 minutes, out of which 27 minutes were spent doing final gathering. There were 15,200 final gathers, with 11.7 million final gather rays. Doing the final gathering without precomputed irradiances

(a)   (b)

(c)   (d)

(e)

Figure 2. Interior: (a) Global photon map. (b) Irradiance estimates computed precisely at the image sample points (dimmed for display). (c) Precomputed irradiance estimates at all 500,000 photon positions (dimmed for display). (d) Radiance estimates based on (c). (e) Complete image with direct illumination, specular reflection, and soft indirect illumination. (See Color Plate II.)

takes 169 minutes, so the final gathering time is reduced by a factor of 6.3. The 6 minutes spent on precomputation corresponds to 4.2% of the 142 minutes saved during rendering.

Acknowledgments. Thanks to Hoang-My Christensen, Pete Shirley, and the anonymous reviewer for many helpful suggestions to improve the structure, contents, and readability of this paper. Thanks to Dani Lischinski for providing the interior scene in Figure 2. It was originally created by Matt Hyatt at the Cornell Program of Computer Graphics, and later extended by Eric Stollnitz.

References

[Bentley 75] Jon L. Bentley. "Multidimensional Binary Search Trees Used for Associative Searching." *Communications of the ACM,* 18(9): 509–517 (September 1975).

[de Berg et al. 97] Mark de Berg, Marc van Kreveld, Mark Overmars, and Otfried Schwarzkopf. *Computational Geometry—Algorithms and Applications.* Berlin: Springer-Verlag, 1997.

[Christensen et al. 96] Per H. Christensen, Eric J. Stollnitz, David H. Salesin, and Tony D. DeRose. "Global Illumination of Glossy Environments using Wavelets and Importance." *ACM Transactions on Graphics,* 15(1): 37–71 (January 1996).

[Cook et al. 84] Robert L. Cook, Thomas Porter, and Loren Carpenter. "Distributed Ray Tracing." *Computer Graphics (Proc. SIGGRAPH 84),* 18(3): 137–145 (July 1984).

[Glassner 89] Andrew S. Glassner. *An Introduction to Ray Tracing.* London: Academic Press, 1989.

[Glassner 95] Andrew S. Glassner. *Principles of Digital Image Synthesis.* San Francisco: Morgan Kaufmann Publishers, 1995.

[Jensen 95] Henrik Wann Jensen. "Importance Driven Path Tracing using the Photon Map." In *Rendering Techniques '95 (Proceedings of the 6th Eurographics Workshop on Rendering),* pp. 326–335. Vienna: Springer-Verlag, 1995.

[Jensen 96a] Henrik Wann Jensen. "Global Illumination using Photon Maps." In *Rendering Techniques '96 (Proceedings of the 7th Eurographics Workshop on Rendering),* pp. 21–30. Vienna: Springer-Verlag, 1996.

[Jensen 96b] Henrik Wann Jensen. "Rendering Caustics on Non-Lambertian Surfaces." In *Proceedings of Graphics Interface '96,* pp. 116-121. San Francisco: Morgan Kaufmann Publishers, May 1996. (Later published in *Computer Graphics Forum,* 16(1): 57–64 (March 1997).)

[Jensen 96c] Henrik Wann Jensen. *The Photon Map in Global Illumination.* Ph.D. dissertation, Technical University of Denmark, September 1996.

Vol. 4, No. 3: 1–10   journal of graphics tools

[Jensen, Christensen 98] Henrik Wann Jensen and Per H. Christensen. "Efficient Simulation of Light Transport in Scenes with Participating Media using Photon Maps." *Computer Graphics (Proc. SIGGRAPH 98)*, pp. 311–320 (July 1998).

[Reichert 92] Mark C. Reichert. *A Two-Pass Radiosity Method Driven by Lights and Viewer Position.* Master's thesis, Cornell University, January 1992.

[Smits et al. 92] Brian E. Smits, James R. Arvo, and David H. Salesin. "An Importance-Driven Radiosity Algorithm." *Computer Graphics (Proc. SIG-GRAPH 92)* 26(2): 273–282 (July 1992).

[Ward 91] Gregory Ward. "Real Pixels." In *Graphics Gems II,* edited by James R. Arvo, pp. 80–83. Cambridge, MA: Academic Press, 1991.

[Ward, Heckbert 92] Gregory Ward and Paul Heckbert. "Irradiance Gradients." In *Proceedings of the 3rd Eurographics Workshop on Rendering,* pp. 85–98. Eurographics, May 1992.

[Ward, Shakespeare 98] Gregory Ward Larson and Rob Shakespeare. *Rendering with Radiance—The Art and Science of Lighting Visualization.* San Francisco: Morgan Kaufmann Publishers, 1998.

[Whitted 80] Turner Whitted. "An Improved Illumination Model for Shaded Display." *Communications of the ACM,* 23(6): 343–349 (June 1980).

Web Information:

The images are available on the web at
http://www.acm.org/jgt/papers/Christensen99

Per H. Christensen, Square USA, 55 Merchant Street #3100, Honolulu, HI 96813.
(per.christensen@acm.org)

Received June 24, 1999; accepted December 22, 1999.

New Since Original Publication

There are numerous variations and extensions of the method presented in this paper.

We can compute the irradiances "on demand" instead of as a preprocess. When a final gather ray hits a surface point, we find the nearest photon with appropriate normal. If that photon has an irradiance value stored with it, we simply use it. If the photon does not have an irradiance value, we find the nearest n photons, compute the irradiance, store it with that photon, and use that value. The advantage of computing the irradiance on demand like this is that if some photons are in parts of the scene that are never seen (neither directly nor indirectly) in the image, then their irradiance will not be computed.

Another variation is to store radiance instead of irradiance with the photons. This is done by multiplying the irradiance estimate by the appropriate diffuse reflection coefficients. Ideally, the diffuse reflection coefficients must be determined with a texture mip map look-up at an appropriate level, corresponding to the area of the Voronoi cell for which the radiance is determined. This is of course an even coarser approximation of the illumination of the scene: in addition to approximating the irradiance as piece-wise constant, we also approximate the textures as piece-wise constant. However, this approximation is fully sufficient for final gathering, and the smoothing actually makes it faster to sample the indirect illumination since a smooth function can be sampled with fewer samples than a function with high frequency details. The advantage of storing radiance is that it reduces the number of shader evaluations and texture map lookups during rendering—which in some cases can make a significant difference in render times.

Other variations and extensions are discussed in the chapter "Photon Mapping Tricks" in *A Practical Guide to Global Illumination using Photon Mapping*, SIGGRAPH 2001 Course Note Number 38 and SIGGRAPH 2002 Course Note Number 43.

Current Contact Information:

Per H. Christensen, Pixar Animation Studios, 506 Second Avenue, Suite 640, Seattle, WA 98104 (per.christensen@acm.org)

Part VI

Sampling and Shading

Vol. 2, No. 2: 9–24

Sampling with Hammersley and Halton Points

Tien-Tsin Wong
The Chinese University of Hong Kong

Wai-Shing Luk
Katholieke Universiteit Leuven

Pheng-Ann Heng
The Chinese University of Hong Kong

Abstract. The Hammersley and Halton point sets, two well-known, low discrepancy sequences, have been used for quasi-Monte Carlo integration in previous research. A deterministic formula generates a uniformly distributed and stochastic-looking sampling pattern at low computational cost. The Halton point set is also useful for incremental sampling. In this paper, we discuss detailed implementation issues and our experience of choosing suitable bases for the point sets, not just on the two-dimensional plane but also on a spherical surface. The sampling scheme is also applied to ray tracing, with a significant improvement in error over standard sampling techniques.

1. Introduction

Many different sampling techniques are used in computer graphics for the purpose of antialiasing. The two easiest ways to sample are randomly and regularly. Unfortunately, random sampling gives a noisy result. Regular sampling causes aliasing which requires many extra samples to reduce.

Vol. 2, No. 2: 9–24

Several techniques in between random and regular sampling have been proposed. The thesis of Shirley [Shirley 91b] surveyed the common sampling
techniques including jittered [Cook et al. 84], semijittered, Poisson disk,
and N-rooks sampling. Cychosz generated [Cychosz 90] sampling jitters
using look-up tables. Chiu et al. [Chiu et al. 94] combined jittered and
N-rooks methods to design a new multijittered sampling. Cross [Cross 95]
used a genetic algorithm to find the optimal sampling pattern for uniformly
distributed edges. All these methods make trade-offs between noisiness and
aliasing.

A sampling technique is hierarchical if, when requested to generate N_0
samples, the result coincides with the first N_0 samples it would generate in
a sequence of $N = N_0 + 1$ samples. This feature is useful since the number
of samples can be incrementally increased without recalculating the previous
ones. Shoemake [Shoemake 91] mentioned a means to incrementally sample one-dimensional space while keeping the samples as uniform as possible.
However, this method is not easily generalized to higher dimensions. Among
previously mentioned methods, only Poisson disk and random sampling are
hierarchical.

Discrepancy analysis measures sample point equidistribution; that is, measures how uniformly distributed the point set is. Shirley [Shirley 91a] first
applied this measurement to graphics problems. The possible importance of
discrepancy in computer graphics was also discussed by Niederreiter [Nieder-
reiter 92a]. Dobkin et al. [Dobkin, Eppstein 93a], [Dobkin, Mitchell 93b],
[Dobkin et al. 96] proposed various methods to measure the discrepancy of
sampling patterns and to generate these patterns [Dobkin et al. 96]. Heinrich and Keller [Heinrich, Keller 94a], [Heinrich, Keller 94b], [Keller 95] and
Ohbuchi and Aono [Ohbuchi, Aono 96] applied low-discrepancy sequences to
Monte Carlo integration in radiosity applications.

In this paper, we discuss two useful low-discrepancy sequences, namely
Hammersley and Halton. These sequences have been used in numerical
[Paskov, Traub 95], [Traub 96], [Case 95] and graphical [Heinrich, Keller 94a],
[Heinrich, Keller 94b], [Keller 95], [Ohbuchi, Aono 96] applications, with a
significant improvement in terms of error. Previous research mainly focused
on sample generation on the two-dimensional plane, cube, and hypercube.
Recently researchers have found [Cui, Freeden 97] that mapping Hammersley points with base of two to the surface of a sphere also gives uniformly
distributed directional vectors. We discuss the implementation issues and experience in choosing suitable bases of Hammersley and Halton points on the
two-dimensional plane and spherical surface.

The mathematical formulation is briefly described in Section 2. Section 3
compares sampling patterns generated using different bases. Ray-tracing experiments to verify the usefulness of the method are discussed in Section 4.
The Appendix lists C implementations.

2. Hammersley and Halton Points

We first describe the definition of Hammersley and Halton points and then discuss their implementation in detail. For more mathematical specifics, readers are referred to the more mathematically based literature [Niederreiter 92b], [Cui, Freeden 97].

Each nonnegative integer k can be expanded using a prime base p:

$$k = a_0 + a_1 p + a_2 p^2 + \cdots + a_r p^r, \tag{1}$$

where each a_i is an integer in $[0, p-1]$. Now we define a function Φ_p of k by

$$\Phi_p(k) = \frac{a_0}{p} + \frac{a_1}{p^2} + \frac{a_2}{p^3} + \cdots + \frac{a_r}{p^{r+1}}. \tag{2}$$

The sequence of $\Phi_p(k)$, for $k = 0, 1, 2, \ldots$, is called the van der Corput sequence [Tezuka 95].

Let d be the dimension of the space to be sampled. Any sequence $p_1, p_2, \ldots,$ p_{d-1} of prime numbers defines a sequence Φ_{p_1}, Φ_{p_2}, $\ldots$, $\Phi_{p_{d-1}}$ of functions whose corresponding k-th d-dimensional Hammersley point is

$$\left(\frac{k}{n}, \Phi_{p_1}(k), \Phi_{p_2}(k), \ldots, \Phi_{p_{d-1}}(k) \right) \quad \text{for } k = 0, 1, 2, \ldots, n-1. \tag{3}$$

Here $p_1 < p_2 < \cdots < p_{d-1}$ and n is the total number of Hammersley points. To evaluate the function $\Phi_p(k)$, the following algorithm can be used.

$$p' = p \, , \ k' = k \, , \ \Phi = 0$$
$$\text{while } k' > 0 \ \text{do}$$
$$\quad a = k' \bmod p$$
$$\quad \Phi = \Phi + \frac{a}{p'}$$
$$\quad k' = \texttt{int}(\frac{k'}{p})$$
$$\quad p' = p'p$$

where $\texttt{int}(x)$ returns the integer part of x.

The above algorithm has a complexity of $O(\log_p k)$ for evaluating the k-th point. Hence the worst-case bound of the algorithm for generating $(N + 1)$ points is

$$\log_p(1) + \log_p(2) + \cdots + \log_p(N-1) + \log_p(N)$$
$$\leq \ \log_p(N) + \log_p(N) + \cdots + \log_p(N) + \log_p(N)$$
$$= \ N \log_p N.$$

A Pascal implementation of this algorithm can be found in [Halton, Smith 64]. In most computer graphics applications, the dimension of the sampled space is either two or three. In this paper, we focus on the generation of a uniformly distributed point set on the surface of the two-dimensional plane and sphere using Hammersley points. Higher dimensional sets can be similarly generated using Formulas (1)–(3).

2.1.   Points on the Two-Dimensional Plane

On the two-dimensional plane, Formula (3) simplifies to

$$\left(\frac{k}{n}, \Phi_{p_1}(k)\right) \quad \text{for } k = 0, 1, 2, \dots, n-1. \tag{4}$$

The range of $\frac{k}{n}$ is $[0, 1)$, while that of $\Phi_{p_1}(k)$ is $[0, 1]$. For computer applications, a good choice of the prime p_1 is $p_1 = 2$. The evaluation of $\Phi_2(k)$ can be done efficiently with about $\log_2(k)$ bitwise shifts, multiplications, and additions: no division is necessary. The C implementation of two-dimensional Hammersley points with base two is shown in the Appendix (Source Code 1). We shift $\frac{k}{n}$ by 0.5 to center the sequence. Otherwise, $\Phi_{p_1}(0)$ will always equal zero for any n, which is a undesirable effect.

However, the original Hammersley algorithm is not hierarchical, due to the first coordinate $\frac{k}{n}$, which for different values of n results in different sets of points. This problem can be resolved by using two p-adic van der Corput sequences with different prime numbers p_1 and p_2. This hierarchical version is known as the Halton point set [Niederreiter 92b], [Tezuka 95].

$$(\Phi_{p_1}(k), \Phi_{p_2}(k)) \quad \text{for } k = 0, 1, 2, \dots, n-1. \tag{5}$$

Since both functions $\Phi_{p_1}(k)$ and $\Phi_{p_2}(k)$ are hierarchical (being independent of n by construction), the Halton point sets are hierarchical as well. Source Code 2 in the Appendix implements the Halton point sets on the two-dimensional plane.

2.2.   Points on the Sphere

To generate directional vectors, or (equivalently) points on the spherical surface, the following mappings [Spanier, Gelbard 69] are needed:

$$\left(\frac{k}{n}, \Phi_p(k)\right) \mapsto (\phi, t) \mapsto \left(\sqrt{1 - t^2} \cos \phi, \sqrt{1 - t^2} \sin \phi, t\right)^T. \tag{6}$$

The first, from $\left(\frac{k}{n}, \Phi_p(k)\right)$ to (ϕ, t), is simply a linear scaling to the required cylindrical domain, $(\phi, t) \in [0, 2\pi) \times [-1, 1]$. The mapping from (ϕ, t) to $(\sqrt{1 - t^2} \cos \phi, \sqrt{1 - t^2} \sin \phi, t)^T$ is a z-preserving radial projection from the unit cylinder $C = \{(x, y, z) \mid x^2 + y^2 = 1 \; |z| \le 1\}$ to the unit sphere.

As before, the coordinate $\frac{k}{n}$ makes the scheme nonhierarchical. Halton points on the sphere can be generated in a similar manner by using two p-adic van der Corput sequences with different prime bases.

$$(\Phi_{p_1}(k), \Phi_{p_2}(k)) \mapsto (\phi, t). \tag{7}$$

Source Code 3 in the Appendix shows the C implementation of Hammersley points on the sphere, with a similar 0.5 shift applied to prevent a fixed sample point from appearing at the south pole. Source Code 4 shows the Halton point version. For efficiency of computation, we fixed $p_1 = 2$ while leaving p_2 as a user input. This restriction can be trivially removed.

3. Appearance

Figures 1 and 2 show the Hammersley points with different bases, on the plane and sphere respectively. We generated 500 samples for the planar test and 1000 for the spherical test. Figures 1(a) and 2(a) are the patterns of random sampling on the plane and sphere respectively. Compared to the random sampling pattern (Figure 1(a)), the Hammersley point set with $p_1 = 2$ (Figure 1(b)) gives a pleasant, less clumped pattern. The points are uniformly distributed without a perceptible pattern. Among the patterns with different bases, a Hammersley point set with $p_1 = 2$ (Figure 1(b)) also gives the most uniformly distributed pattern. As the base p_1 increases (from Figures 1(b)–1(f)), the pattern becomes more and more regular. The points tend to line up in slanting lines, which will clearly increase aliasing problems.

The same progression affects spherical sampling patterns (Figures 2(b)–2(f)). When $p_1 = 2$, it gives the best uniformly distributed pattern on the sphere. Cui et al. [Cui, Freeden 97] measured the uniformity of Hammersley points with $p_1 = 2$ on the sphere using the generalized discrepancy. Hammersley points with $p_1 = 2$ give the lowest generalized discrepancy (most uniformly distributed) among the methods tested. As p_1 increases (from Figures 2(b)–2(f)), points start to line up and form regular lines on the sphere. The position of the pole (marked with an arrow) becomes distinguishable from the pattern.

The Halton point sets give patterns with varying uniformity and regularity (Figures 3 and 4). To compare the effect of different bases p_1 and p_2, all patterns generated with $p_1 = 2$ are placed on the left, and those with $p_1 = 3$ on the right. The omission of the case where $p_1 = p_2 = 3$ is due to the constraint $p_1 < p_2$. Figure 3(b) gives a pattern with somewhat aligned points. Others give rather pleasant appearances. Among the point sets tested, none gives a better pattern than Hammersley points with $p_1 = 2$. In general, the patterns of Halton points are quite unpredictable. Nevertheless, after transforming the points to the sphere, the pole and equator of the sphere become indistinguishable (Figures 4(a)–4(e)). They are not as uniformly distributed as the Hammersley point set with $p_1 = 2$, but no lining-up occurs like that observed in Hammersley points.

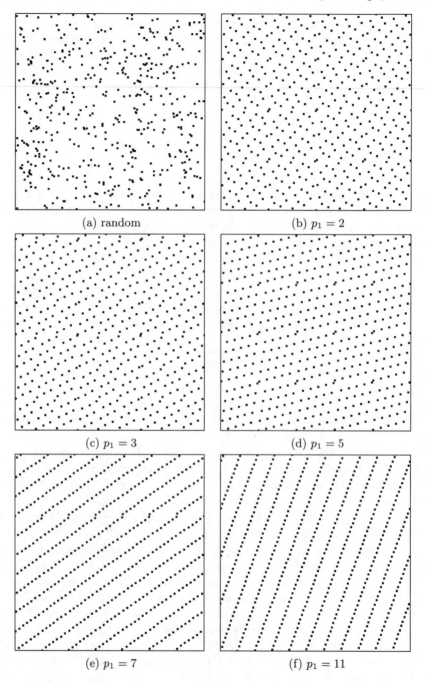

(a) random

(b) $p_1 = 2$

(c) $p_1 = 3$

(d) $p_1 = 5$

(e) $p_1 = 7$

(f) $p_1 = 11$

Figure 1. Hammersley points on the two-dimensional plane ($n = 500$).

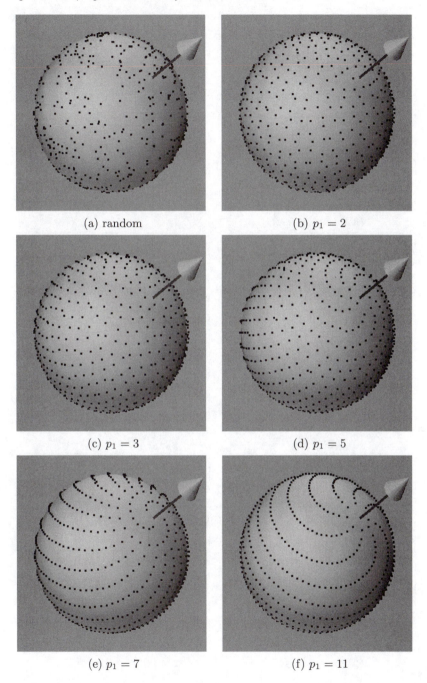

(a) random

(b) $p_1 = 2$

(c) $p_1 = 3$

(d) $p_1 = 5$

(e) $p_1 = 7$

(f) $p_1 = 11$

Figure 2. Hammersley points on the sphere ($n = 1000$).

Vol. 2, No. 2: 9–24

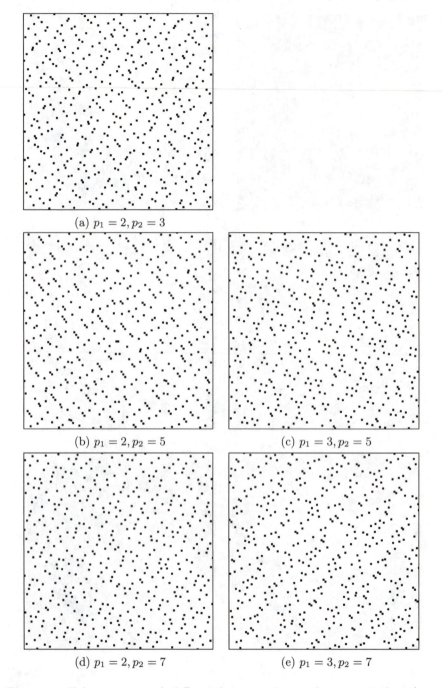

(a) $p_1 = 2, p_2 = 3$

(b) $p_1 = 2, p_2 = 5$ (c) $p_1 = 3, p_2 = 5$

(d) $p_1 = 2, p_2 = 7$ (e) $p_1 = 3, p_2 = 7$

Figure 3. Halton points with different bases on the two-dimensional plane ($n = 500$).

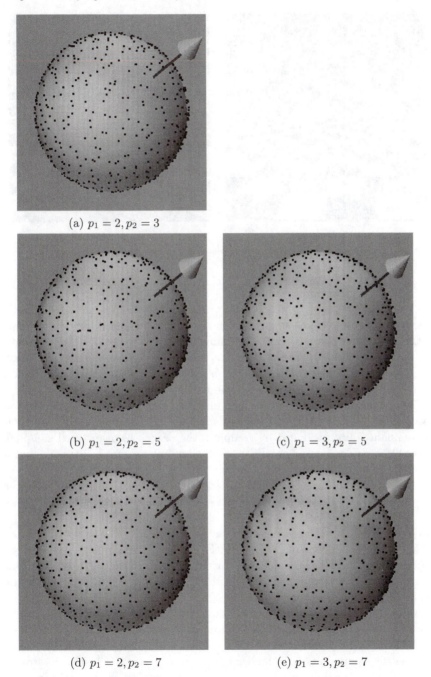

(a) $p_1 = 2, p_2 = 3$

(b) $p_1 = 2, p_2 = 5$

(c) $p_1 = 3, p_2 = 5$

(d) $p_1 = 2, p_2 = 7$

(e) $p_1 = 3, p_2 = 7$

Figure 4. Halton points with different bases on the sphere ($n = 1000$).

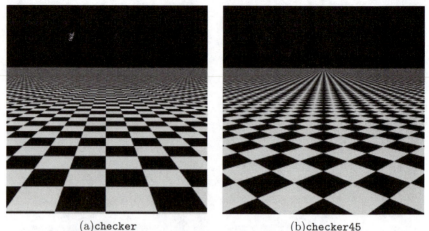

(a)checker   (b)checker45

Figure 5. The two test scenes used in the sampling test.

4. Ray-Tracing Experiments

The method is tested in a ray tracer. Instead of generating a distinct sampling pattern for each pixel, a single sampling pattern is generated for the whole screen. Otherwise, the sampling pattern for each pixel would be the same, since the Hammersley and Halton points are actually deterministic. Hence, we can only specify the *average* sample per pixel.

Two scenes are chosen for testing: checker (Figure 5(a)) and checker45 (Figure 5(b)). The "correct" images, used for calculating the pixel error E in luminance, are produced by sampling the scenes using jittered sampling with 400 samples per pixel. Five other sampling schemes—jittered, multi-jittered, Poisson disk, random, and regular—are included for comparison. All of these five sampling schemes are tested with 16 samples per pixel, while the Hammersley and Halton point sets are tested with an average of 16 samples per pixel.

Four statistical data are recorded: average (Mean$|E|$), standard deviation (S.D.($|E|$)), root-mean-square (R.M.S.(E)), and maximum (Max.($|E|$)) of the absolute pixel error in luminance. Tables 1 and 2 show the statistics from test scenes checker and checker45 respectively. Methods listed in the tables are ranked by their performance.

Among the tested methods, the Hammersley point set with $p_1 = 2$ gives the lowest average, standard derivation, and root-mean-square of absolute pixel errors in both test scenes. Multijittered sampling is the first runner-up. Hammersley point sets with higher bases ($p_1 > 3$) are not tested due to the lining-up phenomenon, which certainly introduces aliasing. For Halton point sets, we arbitrarily choose two bases for testing since there is no general trend

Methods	Mean($|E|$)	S.D.($|E|$)	R.M.S.(E)	Max.($|E|$)
Hamm., $p_1 = 2$	0.0086	0.0247	0.0261	0.3451
multi-jitter, $n = 4, N = 16$	0.0091	0.0261	0.0277	0.3843
Hamm., $p_1 = 3$	0.0097	0.0265	0.0282	0.3961
Halton, $p_1 = 2, p_2 = 7$	0.0105	0.0280	0.0299	0.3451
Halton, $p_1 = 2, p_2 = 3$	0.0110	0.0291	0.0312	0.3686
jittered, 4×4	0.0128	0.0335	0.0358	0.3804
Poisson, $d = 0.2$	0.0132	0.0338	0.0363	0.3804
random	0.0179	0.0443	0.0478	0.3961
regular	0.0188	0.0491	0.0526	0.5098

Table 1. Statistics of the ray-traced image **checker**. E is the pixel error in luminance.

Methods	Mean($|E|$)	S.D.($|E|$)	R.M.S.(E)	Max.($|E|$)
Hamm., $p_1 = 2$	0.0101	0.0264	0.0282	0.3882
multi-jitter, $n = 4, N = 16$	0.0103	0.0270	0.0289	0.3686
Hamm., $p_1 = 3$	0.0106	0.0274	0.0294	0.4431
Halton, $p_1 = 2, p_2 = 3$	0.0114	0.0287	0.0309	0.3882
Halton, $p_1 = 2, p_2 = 7$	0.0131	0.0289	0.0310	0.4118
jittered, 4×4	0.0131	0.0332	0.0357	0.3765
Poisson, $d = 0.2$	0.0133	0.0332	0.0358	0.4118
regular	0.0138	0.0393	0.0416	0.5059
random	0.0185	0.0446	0.0483	0.4000

Table 2. Statistics of the ray-traced image **checker45**. E is the pixel error in luminance.

in the appearance of the patterns. In our experiment, Hammersley point sets are better than the tested Halton point sets. Both Hammersley and Halton point sets give lower error than that of traditional jittered and Poisson disk sampling except the multijittered method.

5. Conclusion

The Hammersley point set with $p_1 = 2$ gives the most uniformly distributed sampling pattern. For higher p_1, the points tend to align and reduce its usefulness. Although the Halton point sets do not give patterns as uniformly distributed as Hammersley point sets, they do not have the line-up problem and allow incremental sampling.

Hammersley points and Halton points have proved useful for quasi-Monte Carlo integration. The methods have been applied to ray-tracing applications with a significant improvement in pixel error. The complexity of both the Hammersley and Halton point generation algorithms is $O(N \log_p N)$, which is smaller than that of Poisson disk.

Acknowledgments. We would like to thank Prof. Timothy Poston of National University of Singapore for his careful proofread and useful suggestions. We would also like to express our thanks to the editor and reviewers for their valuable advice and comments.

Appendix: Source Code

Source Code 1. *Hammersley Points on Two-Dimensional Plane with* $p_1 = 2$

```
void PlaneHammersley(float *result, int n)
{
  float p, u, v;
  int k, kk, pos;

  for (k=0, pos=0 ; k<n ; k++)
  {
    u = 0;
    for (p=0.5, kk=k ; kk ; p*=0.5, kk>>=1)
      if (kk & 1)                          // kk mod 2 == 1
        u += p;
    v = (k + 0.5) / n;
    result[pos++] = u;
    result[pos++] = v;
  }
}
```

Source Code 2. *Halton Points on Two-Dimensional Plane with* $p_1 = 2$

```
void PlaneHalton(float *result, int n, int p2)
{
  float p, u, v, ip;
  int k, kk, pos, a;

  for (k=0, pos=0 ; k<n ; k++)
  {
    u = 0;
    for (p=0.5, kk=k ; kk ; p*=0.5, kk>>=1)
      if (kk & 1)                          // kk mod 2 == 1
        u += p;
    v = 0;
    ip = 1.0/p2;                           // inverse of p2
    for (p=ip, kk=k ; kk ; p*=ip, kk/=p2)  // kk = (int)(kk/p2)
      if ((a = kk % p2))
        v += a * p;
    result[pos++] = u;
    result[pos++] = v;
  }
}
```

Source Code 3. *Hammersley Points on Sphere with $p_1 = 2$*

```
void SphereHammersley(float *result, int n)
{
  float p, t, st, phi, phirad;
  int k, kk, pos;

  for (k=0, pos=0 ; k<n ; k++)
  {
    t = 0;
    for (p=0.5, kk=k ; kk ; p*=0.5, kk>>=1)
      if (kk & 1)                          // kk mod 2 == 1
        t += p;
    t = 2.0 * t  - 1.0;                     // map from [0,1] to [-1,1]
    phi = (k + 0.5) / n;                    // a slight shift
    phirad =  phi * 2.0 * M_PI;             // map to [0, 2 pi)
    st = sqrt(1.0-t*t);
    result[pos++] = st * cos(phirad);
    result[pos++] = st * sin(phirad);
    result[pos++] = t;
  }
}
```

Source Code 4. *Halton Points on Sphere with $p_1 = 2$*

```
void SphereHalton(float *result, int n, int p2)
{
  float p, t, st, phi, phirad, ip;
  int k, kk, pos, a;

  for (k=0, pos=0 ; k<n ; k++)
  {
    t = 0;
    for (p=0.5, kk=k ; kk ; p*=0.5, kk>>=1)
      if (kk & 1)                          // kk mod 2 == 1
        t += p;
    t = 2.0 * t - 1.0;                      // map from [0,1] to [-1,1]
    st = sqrt(1.0-t*t);
    phi = 0;
    ip = 1.0/p2;                            // inverse of p2
    for (p=ip, kk=k ; kk ; p*=ip, kk/=p2)// kk = (int)(kk/p2)
      if ((a = kk % p2))
        phi += a * p;
    phirad =  phi * 4.0 * M_PI;             // map from [0,0.5] to [0, 2 pi)
    result[pos++] = st * cos(phirad);
    result[pos++] = st * sin(phirad);
    result[pos++] = t;
  }
}
```

References

[Case 95] James Case. "Wall Street's Dalliance with Number Theory." *SIAM News* (December 1995).

[Chiu et al. 94] Kenneth Chiu, Peter Shirley, and Changyaw Wang. "Multi-Jittered Sampling." In *Graphics Gems IV*, edited by Paul S. Heckbert, p. 370–374. Cambridge, MA: AP Professional, 1994.

[Cook et al. 84] Robert L. Cook, Thomas Porter, and Loren Carpenter. "Distributed Ray-Tracing." *Computer Graphics (Proc. SIGGRAPH 84)*, 18(3):137–145(July 1984).

[Cross 95] Robert A. Cross. "Sampling Patterns Optimized for Uniform Distribution of Edges." In *Graphics Gems V*, edited by Alan W. Paeth, p. 359–363. Chestnut Hill, MA: AP Professional, 1995.

[Cui, Freeden 97] Jianjun Cui and Willi Freeden. "Equidistribution on the Sphere." *SIAM Journal on Scientific Computing*, 18(2):595–609(March 1997).

[Cychosz 90] Joseph M. Cychosz. "Efficient Generation of Sampling Jitter Using Look-Up Tables." In *Graphics Gems*, edited by Andrew Glassner, p. 64–74. Cambridge, MA: AP Professional, 1990.

[Dobkin, Eppstein 93a] D. P. Dobkin and D. Eppstein. "Computing the Discrepancy." In *Proceedings of the 9th ACM Symposium on Computational Geometry*, p. 47–52. New York: ACM, 1993.

[Dobkin et al. 96] D. P. Dobkin, David Eppstein, and Don P. Mitchell. "Computing the Discrepancy with Applications to Supersampling Patterns." *ACM Transactions on Graphics*, 15(4):354–376(October 1996).

[Dobkin, Mitchell 93b] D. P. Dobkin and D. P. Mitchell. "Random-Edge Discrepancy of Supersampling Patterns." In *Graphics Interface*, p. 62–69. Toronto: Canadian Information Processing Society, 1993.

[Halton, Smith 64] J. H. Halton and G. B. Smith. "Radical-Inverse Quasi-Random Point Sequence." *Communications of the ACM*, 7(12):701–702(December 1964).

[Heinrich, Keller 94a] Stefan Heinrich and Alexander Keller. "Quasi-Monte Carlo Methods in Computer Graphics, Part I: The qmc Buffer." Technical Report 242/94, University of Kaiserlautern, 1994.

[Heinrich, Keller 94b] Stefan Heinrich and Alexander Keller. "Quasi-Monte Carlo Methods in Computer Graphics, Part II: The Radiance Equation." Technical Report 243/94, University of Kaiserlautern, 1994.

[Keller 95] Alexander Keller. "A Quasi-Monte Carlo Algorithm for the Global Illumination Problem in the Radiosity Setting." In *Proceedings of Monte Carlo and Quasi-Monte Carlo Methods in Scientific Computing*, p. 239–251. New York: Springer-Verlag, 1995.

[Niederreiter 92a] H. Niederreiter. "Quasirandom Sampling Computer Graphics." In *Proceedings of the 3rd International Seminar on Digital Image Processing in Medicine*, p. 29–33. Riga: Latvian Academy of Sciences, 1992.

[Niederreiter 92b] H. Niederreiter. *Random Number generation and Quasi-Monte Carlo Methods*. Philadephia: CBMS-NSF and SIAM, 1992.

[Ohbuchi, Aono 96] Ryutarou Ohbuchi and Masaki Aono. "Quasi-Monte Carlo Rendering with Adaptive Sampling." Technical Report, Tokyo Research Laboratory, IBM Japan Ltd., 1996.

[Paskov, Traub 95] S. H. Paskov and J. F. Traub. "Faster Valuing of Financial Derivatives." *Journal of Portfolio Management*, 22:113–120(1995).

[Shirley 91a] Peter Shirley. "Discrepancy as a Quality Measure for Sample Distributions." In *Proceedings of Eurographics*, edited by Werner Purgathofer, p. 183–193. Amsterdam: North-Holland, 1991.

[Shirley 91b] Peter Shirley. "Physically Based Lighting Calculations for Computer Graphics." PhD Thesis, University of Illinois at Urbana-Champaign, 1991.

[Shoemake 91] Ken Shoemake. "Interval Sampling." In *Graphics Gems II*, edited by James Arvo, p. 394–395. San Diego, CA: AP Professional, 1991.

[Spanier, Gelbard 69] Jerome Spanier and Ely M. Gelbard. "Monte Carlo Principles and Neutron Transport Problems." New York, Addison-Wesley, 1969.

[Tezuka 95] Shu Tezuka. "Uniform Random Numbers: Theory and Practice." Boston: Kluwer Academic Publishers, 1995.

[Traub 96] Joseph Traub. "In Math We Trust." *What's Happening in the Mathematical Sciences*, 3:101–111(1996).

Web Information:

All source codes in the Appendix and a demonstration program showing the appearances of various Hammersley and Halton point sets are available at
http://www.acm.org/jgt/papers/WongLukHeng97.

Tien-Tsin Wong, Department of Computer Science and Engineering, The Chinese University of Hong Kong, Shatin, Hong Kong (ttwong@acm.org)

Vol. 2, No. 2: 9–24   journal of graphics tools

Wai-Shing Luk, Departement Computerwetenschappen, Katholieke Universiteit Leuven, Celestijnenlaan 200A, B-3001 Heverlee, Belgium (Wai-Shing.Luk@cs.kuleuven.ac.be)

Pheng-Ann Heng, Department of Computer Science and Engineering, The Chinese University of Hong Kong, Shatin, Hong Kong (pheng@cse.cuhk.edu.hk)

Received February 6, 1997; accepted April 23, 1997

Current Contact Information:

Tien-Tsin Wong, Department of Computer Science and Engineering, The Chinese University of Hong Kong, Shatin, Hong Kong (ttwong@cse.cuhk.edu.hk)

Wai-Shing Luk, Departement Computerwetenschappen, Katholieke Universiteit Leuven, Celestijnenlaan 200A, B-3001 Heverlee, Belgium (Wai-Shing.Luk@cs.kuleuven.ac.be)

Pheng-Ann Heng, Department of Computer Science and Engineering, The Chinese University of Hong Kong, Shatin, Hong Kong (pheng@cse.cuhk.edu.hk)

Vol. 2, No. 3: 45–52

A Low Distortion Map Between Disk and Square

Peter Shirley
University of Utah

Kenneth Chiu
Indiana University

Abstract. This paper presents a map between squares and disks that associates concentric squares with concentric circles. This map preserves adjacency and fractional area, and has proven useful in many sampling applications where correspondences must be maintained between the two shapes. The paper also provides code to compute the map that minimizes branching and is robust for all inputs. Finally, it extends the map to the hemisphere. Though this map has been used in publications before, details of its computation have never previously been published.

1. Introduction

Many graphics applications map points on the unit square $\mathcal{S} = [0,1]^2$ to points on the unit disk $\mathcal{D} = \{(x,y) \,|\, x^2 + y^2 \leq 1\}$. Sampling disk-shaped camera lenses is one such application. Though each application can have its own unique requirements, experience has shown that a good general-purpose map should have the following properties:

- *Preserves fractional area.* Let $R \in \mathcal{S}$ be a region in domain $\mathcal{S}$. The fractional area of R is defined as $a(R)/a(\mathcal{S})$, where the function a denotes area. Then a map $m : \mathcal{S} \to \mathcal{D}$ preserves fractional area if the fractional area of R is the same as the fractional area $a(m(R))/a(\mathcal{D})$ for all $R \in \mathcal{S}$ (see Figure 1). This property ensures that a "fair" set of points on the

Vol. 2, No. 3: 45–52

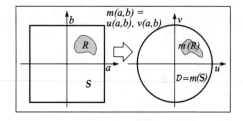

Figure 1. A map that preserves fractional area will map regions with the area constraint that $a(A_1)/a(\mathcal{S}) = a(A_2)/a(\mathcal{D})$.

square will map to a fair set on the disk. For distribution ray tracers, this means that a jittered set of points on the square will transform to a jittered set of points on the disk.

- *Bicontinuous.* A map is *bicontinuous* if the map and its inverse are both continuous. Such a map will preserve adjacency; that is, points that are close on the disk come from points that are close on the square, and vice versa. This is especially useful when working on the hemisphere, because it means that the angular distance between two directions can be estimated by their linear distance on the square.

- *Low distortion.* Low distortion means that shapes are reasonably well-preserved. Defining distortion more formally is possible, but probably of limited benefit since no single definition is clearly suitable for a wide range of applications.

These properties, as well as several others, are also important to map-makers who map the spherical earth to a rectangular map [Canters, Decleir 89]. Figure 2 shows the *polar map,* probably the most obvious way to map the square to the disk: x is mapped to r and y is mapped to ϕ. Because the area between r and $r + \Delta r$ is proportional to r to the first-order, the map is not a linear function of x:

$$r = \sqrt{x}$$
$$\phi = 2\pi y$$

This map preserves fractional area but does not satisfy the other properties. As can be seen in the image, shapes are grossly distorted. Another problem is that although $(r, \phi) = (\sqrt{x}, 2\pi y)$ is continuous, the inverse map is not. For some applications, a bicontinuous map is preferable.

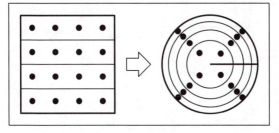

Figure 2. The polar map takes horizontal strips to concentric rings.

2.   The Concentric Map

In this section, the *concentric map* is presented, which maps concentric squares to concentric circles (shown in Figure 3). This map has the properties listed above, and it is easy to compute. It has been advocated for ray-tracing applications [Shirley 90] and has been shown empirically to provide lower error for stochastic camera-lens sampling [Kolb et al. 95], but the details of its computation have not been previously published.

The algebra behind the map is shown in Figure 4. The square is mapped to $(a, b) \in [-1, 1]^2$, and is divided into four regions by the lines $a = b$ and $a = -b$. For the first region, the map is:

$$
\begin{aligned}
r &= a \\
\phi &= \frac{\pi}{4}\frac{b}{a}
\end{aligned}
$$

This produces an angle $\phi \in [-\pi/4, \pi/4]$. The other four regions have analogous transforms. Figure 5 shows how the polar map and concentric map transform the connected points in a uniform grid. The Appendix shows that this map preserves fractional area.

This concentric square to concentric circle map can be implemented in a small piece of robust code written to minimize branching:

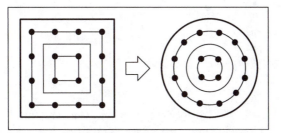

Figure 3. The concentric map takes concentric square strips to concentric rings.

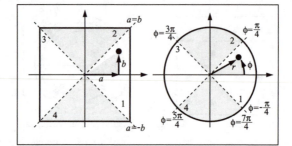

Figure 4. Quantities for mapping region 1.

```
Vector2 ToUnitDisk( Vector2 onSquare )
  real phi, r, u, v
  real a = 2*onSquare.X-1      // (a,b) is now on [-1,1]^2
  real b = 2*onSquare.Y-1

  if (a > -b)                  // region 1 or 2
    if (a > b)                 // region 1, also |a| > |b|
        r = a
        phi = (PI/4 ) * (b/a)
    else                       // region 2, also |b| > |a|
        r = b;
        phi = (PI/4) * (2 - (a/b))
  else                         // region 3 or 4
    if (a < b)                 // region 3, also |a| >= |b|, a != 0
      r = -a
      phi = (PI/4) * (4 + (b/a))
    else // region 4, |b| >= |a|, but a==0 and b==0 could occur.
      r = -b
      if (b != 0)
          phi = (PI/4) * (6 - (a/b))
      else
          phi = 0

  u = r * cos( phi )
  v = r * sin( phi )
  return Vector2( u, v )
end
```

The inverse map is straightforward provided that the function atan2() is available. The code below assumes atan2() returns a number in the range $[-\pi, \pi]$, as it does in ANSI C, Draft ANSI C++, ANSI *FORTRAN*, and *Java*:

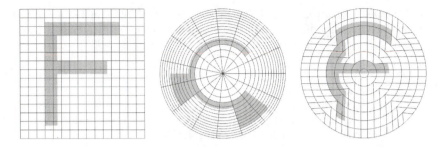

Figure 5. Left: Unit square with gridlines and letter "F". Middle: Polar mapping of gridlines and "F". Right: Concentric mapping of gridlines and "F".

```
Vector2 FromUnitDisk(Vector2 onDisk)
    real r = sqrt(onDisk.X * onDisk.X + onDisk.Y * onDisk.Y)
    real phi = atan2(onDisk.Y, onDisk.X)
    if (phi < -PI/4) phi += 2*PI      // in range [-pi/4,7pi/4]
    real a, b, x, y
    if (phi < PI/4)                   // region 1
        a = r
        b = phi * a / (PI/4)
    else if (phi < 3*PI/4 )           // region 2
        b = r
        a = -(phi - PI/2) * b / {PI/4)
    else if (phi <  5*PI/4 )          // region 3
        a = -r
        b = (phi - PI) * a / (PI/4)
    else                              // region 4
        b = -r
        a = -(phi - 3*PI/2) * b / (PI/4)
    x = (a + 1) / 2
    y = (b + 1) / 2
    return Vector2(x, y)
end
```

The concentric map can be extended to the hemisphere by projecting the points up from the disk to the hemisphere. By controlling exactly how the points are projected, different distributions are generated, as shown in the next section.

3. Applications

The concentric map can be used in a number of different ways:

- *Random point on a disk.* A random point on the disk of radius r can be generated by passing a random point on $[0,1]^2$ through the map and multiplying both coordinates by r.

- *Jittered point on a disk.* A set of n^2 jittered points can be generated similarly using uniform random numbers between 0 and 1, such as those generated by a call to `drand48()`:

```
int s = 0;
for (int i = 0; i < n; i++)
   for (int j = 0; j < n; j++)
      Vector2 onSquare((i + drand48()) / n, (j + drand48()) / n )
      onDisk[s++] = ToUnitDisk(onSquare )
```

- *Cosine distribution on a hemisphere.* Radiosity and path-tracing applications often need points on the unit hemisphere with density proportional to the z-coordinate of the point. This is commonly referred to as a *cosine distribution,* because the z-coordinate of a point on a unit sphere is the cosine between the z-axis and the direction from the sphere's center to the point. Such a distribution is commonly generated from uniform points on the disk by projecting the points along the z-axis. Assuming that (u, v) are the coordinates on the disk and letting $r = \sqrt{u^2 + v^2}$, the cosine-distributed point (x, y, z) on the hemisphere oriented along the positive z-axis is

$$
\begin{aligned}
x &= u \\
y &= v \\
z &= \sqrt{1 - r^2}
\end{aligned}
$$

- *Uniform distribution on a hemisphere.* Generating a uniform distribution of the z-coordinate will create a uniform distribution on the hemisphere. This can be accomplished by noting that if there is a uniform distribution on a disk, then the distribution of r^2 is also uniform. So given a uniformly distributed random point (u, v) on a disk, a uniformly distributed point is produced on the hemisphere by first generating a z-coordinate from r^2, and then assigning x and y such that the point is on the hemisphere [Shirley 92].

$$
\begin{aligned}
x &= u \frac{\sqrt{1 - z^2}}{r} \\
y &= v \frac{\sqrt{1 - z^2}}{r} \\
z &= 1 - r^2
\end{aligned}
$$

- *Phong-like distribution on a hemisphere.* The uniform and cosine distributions can be generalized to a Phong-like distribution where density is proportional to a power of the cosine, or z^N:

$$x = u\frac{\sqrt{1-z^2}}{r}$$

$$y = v\frac{\sqrt{1-z^2}}{r}$$

$$z = \left(1-r^2\right)^{\frac{1}{N+1}}$$

Note that the two previous formulas are special cases of this formula.

- *Subdivision data.* In any application where a subdivision data structure must be kept on the hemisphere or disk, the bicontinuous and low-distortion properties make the map ideal. For example, a *k-d tree* organizing points on the disk can be kept on the transformed points on the square so that conventional axis-parallel two-dimensional divisions can be used.

Appendix

When the concentric map is expanded out so that $(a, b) \in [-1, 1]^2$ is mapped to (u, v) on the unit disk, the map is (in the first region):

$$u = a\cos\left(\frac{\pi b}{4a}\right)$$

$$v = a\sin\left(\frac{\pi b}{4a}\right)$$

The ratio of differential areas in the range and domain for the map is given by the determinant of the Jacobian matrix. If this determinant is the constant ratio of the area of the entire domain to the entire range, then the map preserves fractional area as desired. This is, in fact, the case for the concentric map:

$$\begin{vmatrix} \frac{\partial u}{\partial a} & \frac{\partial u}{\partial b} \\ \frac{\partial v}{\partial a} & \frac{\partial v}{\partial b} \end{vmatrix} = \begin{vmatrix} \cos\left(\frac{\pi b}{4a}\right) + \frac{\pi b}{4a}\sin\left(\frac{\pi b}{4a}\right) & -\frac{\pi}{4}\sin\left(\frac{\pi b}{4a}\right) \\ \sin\left(\frac{\pi b}{4a}\right) - \frac{\pi b}{4a}\cos\left(\frac{\pi b}{4a}\right) & \frac{\pi}{4}\cos\left(\frac{\pi b}{4a}\right) \end{vmatrix} = \frac{\pi}{4}$$

This value is $\pi/4$ because that is the ratio of the area of the unit disk to the area of $[-1, 1]^2$. By symmetry, the determinant of the Jacobian matrix is the same for the other three regions.

Acknowledgments. Thanks to Eric Lafortune, Peter-Pike Sloan, and Brian Smits for their comments on the paper. Figure 5 was generated using software by Michael Callahan and Nate Robins.

Vol. 2, No. 3: 45–52

References

[Canters, Decleir 89] Frank Canters and Hugo Decleir. *The World in Perspective.* Avon: Bath Press, 1989.

[Kolb et al. 95] Craig Kolb, Pat Hanrahan, and Don Mitchell. "A Realistic Camera Model for Computer Graphics." In *Computer Graphics Proceedings (Proc. SIGGRAPH 95),* edited by Rob Cook, pp. 317–324.

[Shirley 90] Peter Shirley. "A Ray Tracing Method for Illumination Calculation in Diffuse-Specular Scenes." In *Proceedings of Graphics Interface '90,* pp. 205–212. Palo Alto: Morgan Kaufmann, May 1990.

[Shirley 92] Peter Shirley. "Nonuniform Random Point Sets via Warping." In *Graphics Gems III,* edited by David Kirk, pp. 80–83. San Diego: Academic Press, 1992.

Web Information:

C source code is available at
http://www/acm.org/jgt/papers/ShirleyChiu97

Peter Shirley, Computer Science Department, 3190 MEB, University of Utah, Salt Lake City, UT 84112 (shirley@cs.utah.edu)

Kenneth Chiu, Computer Science Department, Lindley Hall, Indiana University, Bloomington, IN 47405 (chiuk@cs.indiana.edu)

Received January 20, 1998; accepted February 17, 1998

Current Contact Information:

Peter Shirley, Computer Science Department, 3190 MEB, University of Utah, Salt Lake City, UT 84112 (shirley@cs.utah.edu)

Kenneth Chiu, Computer Science Department, Lindley Hall, Indiana University, Bloomington, IN 47405 (chiuk@cs.indiana.edu)

Vol. 8, No. 3: 41–47

Generalized Stratified Sampling Using the Hilbert Curve

Mauro Steigleder and Michael D. McCool
University of Waterloo

Abstract. Stratified sampling is a widely used strategy to improve convergence in Monte Carlo techniques. The efficiency of a stratification technique mainly depends on the coherence of the strata. This paper presents an approach to generate an arbitrary number of coherent strata, independently of the dimensionality of the domain, using the Hilbert space-filling curve. Using this approach, it is possible to draw an arbitrary number of stratified samples from higher dimensional spaces using only one-dimensional stratification. This technique can also be used to generalize nonuniform stratified sampling. Source code is available online.

1. Introduction

Many computer graphics problems can be given in an integral form where the integrand is potentially discontinuous and of high dimension. In order to solve these problems, Monte Carlo techniques are often used since their convergence is independent of dimension, and they are robust and easy to implement. However, *pure* Monte Carlo techniques have a slow convergence of $O(N^{-1/2})$. Several techniques have been developed to improve convergence, the most common being *importance sampling*, *stratified sampling*, and *low discrepancy sequences*.

This paper shows how to generalize uniform regular (jittered) stratified sampling to allow the use of an arbitrary number of samples, in domains of any dimension. We also address the integration of this technique with importance

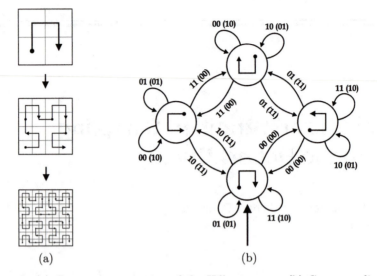

(a)   (b)

Figure 1. (a) Geometric generation of the Hilbert curve. (b) Corresponding automaton to convert between a two-dimensional domain and the unit interval. Arcs are labeled with inputs first followed by outputs, where the input consists of the next high-order bits in x and y coordinates, and the output consists of the next two high-order bits in the result.

sampling using nonuniform sampling. We do not address the integration of this technique with low discrepancy sequences since those techniques deal with higher dimensionality in a different way.

2. Hilbert Space-Filling Curve

Denote the closed unit interval $\mathcal{I} = [0, 1]$ and the closed unit n-dimensional cube $\mathcal{Q} = [0, 1]^n$. A space-filling curve can be informally described as a curve that covers the whole domain $\mathcal{Q}$ without intersecting itself. Formally, a space-filling curve can be represented by a bijective map between $\mathcal{I}$ and $\mathcal{Q}$.

Even though several space-filling curves have been proposed, the Hilbert space-filling curve has properties that make it appropriate for this work. Some of these properties are continuity, compactness, connectivity [Sagan 94], and high coherence [Voorhies 91]. A Hilbert curve can be generated by recursive geometric substitution, as shown in Figure 1(a). When we refer to the Hilbert curve, we specifically refer to the *fractal limit* curve of this rewriting process. In practice, although, we only use a finite number of rewriting steps.

Several algorithms have been proposed to compute the Hilbert bijection [Butz 71, Sagan 92, Warren 03]. We use a simple bit manipulation algorithm

[McCool et al. 01], based on the approach proposed by Bially [Bially 69], to efficiently compute the two-dimensional Hilbert bijection.

Using this approach, the Hilbert rewriting process can be interpreted as a reversible deterministic automaton. Figure 1(b) shows the automaton to convert from the unit square to the unit interval. Arcs are labeled with inputs first, in xy order, followed by outputs. When converting from the unit square to the unit interval, one high-order bit of each coordinate is read at a time, composed into a two-bit symbol, and used as input for the automaton. The automaton then outputs the next two bits of the curve arc length and goes to the next state. Both input and output values are represented as fixed-point numbers. Figure 2 shows the C code for the Hilbert mapping between the unit interval and the two-dimensional square interval. We tabulate the automaton's next-state and output functions and pack them into 32-bit integers for efficiency. This code is available online at the web site listed at the end of this paper.

This approach also works for n-dimensional spaces. In this case, one high-order bit is read from each of the n dimensions, and n high-order bits are output to the result. For spaces with more than two dimensions, we use Butz's algorithm [Butz 71].

When converting from an n-dimensional space to the unit interval, the precision of the result is n times greater than the precision of the input. Conversely, when performing the inverse transform, the precision of the coordinates in the n-dimensional space is n times smaller than the precision of the

```
uint64 Index2Arclength (
   uint32 x,    // input
   uint32 y,    // input
) {
   int t,i,p,s=0;
   uint32 tr=0x8FE65831;
   uint32 out=0x361E9CB4;
   uint64 r;

   for(i=31;i>=0;i--) {
      t = (((x>>i)&1)<<1)|((y>>i)&1);
      p = (s<<3)|(t<<1);
      s = (tr>>p)&3;
      r = (r<<2)|((out>>p)&3);
   }
   return r;
}
```

```
void Arclength2Index (
   uint64 l,    // input
   uint32 *x,   // output
   uint32 *y    // output
) {
   int t,i,p,s=0;
   uint32 tr=0x3E6B94C1;
   uint32 out=0x874B78B4;
   uint32 xx,yy;

   for(i=31;i>=0;i--) {
      t = (l>>(i<<1))&3;
      p = (s<<3)|(t<<1);
      s = (tr>>p)&3;
      xx = (xx<<1)|((out>>p&3)>>1);
      yy = (yy<<1)|((out>>p)&1);
   }
   *x = xx;
   *y = yy;
}
```

(a)   (b)

Figure 2. Code to compute the two-dimensional Hilbert transformations: (a) $\mathcal{Q} \leftrightarrow \mathcal{I}$, and (b) $\mathcal{I} \leftrightarrow \mathcal{Q}$. The state machine of Figure 1(b) is encoded into two 32-bit integers.

curve arc length. For instance, in a two-dimensional space, if the precision in $\mathcal{I}$ is 32 bits, then the maximum precision of the x and y coordinates in $\mathcal{Q}$ is 16 bits. In practice, in order to keep the same precision on higher dimensional spaces, the precision in $\mathcal{I}$ must be increased accordingly.

3. Stratified Sampling Using the Hilbert Curve

When designing a stratification method, it is very important for the strata to be as coherent as possible. As shown by Voorhies [Voorhies 91], the Hilbert curve satisfies this criteria since intervals along the curve cover coherent regions. It can also be shown that two strata in the unit interval with the same length map to regions with the same area [Bially 69].

Denote the Hilbert mapping from the unit interval to the unit n-dimensional cube by $\mathcal{H} : \mathcal{I} \leftrightarrow \mathcal{Q}$. In order to generate m stratified samples in the unit square, we initially generate a set of m stratified samples in the unit interval, $U_m = \{x_1, x_2, \ldots, x_m\}$. Then, applying the Hilbert mapping $\mathcal{H}$ to the set U_m, we obtain a set of m stratified samples in $\mathcal{Q}$. Figure 3 illustrates this process in two dimensions.

The Hilbert space-filling curve can also be used to generate nonuniform stratified samples for nonconstant importance functions. The traditional approach used to draw samples from arbitrary two-dimensional Probability Density Functions (PDFs) uses marginal and conditional probabilities. However, the procedure can be made much simpler by using the Hilbert bijection.

In order to draw a sample from a two-dimensional PDF $p(x_1, x_2)$, we initially reparameterize this PDF to a one-dimensional PDF. Let $x_{\mathcal{H}}$ be the point in $\mathcal{I}$ corresponding to the two-dimensional point (x_1, x_2) in $\mathcal{Q}$. In other

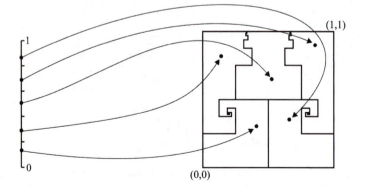

Figure 3. Associated strata in the unit interval and in the unit square.

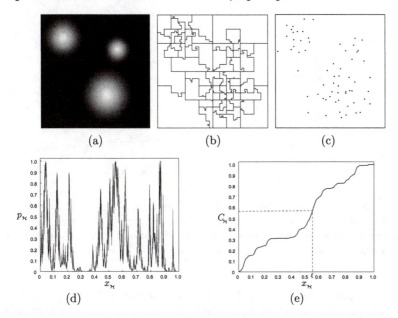

Figure 4. (a) Two-dimensional PDF. (b) Strata distribution using Hilbert stratification. (c) Example of sampling pattern. (d) Reparameterization of the PDF shown in Figure 4(a). (e) Normalized cumulative distribution function of Figure 4(d).

words, let $x_{\mathcal{H}} = \mathcal{H}^{-1}(x_1, x_2)$ and $\mathcal{H}(x_{\mathcal{H}}) = (x_1, x_2)$. The reparameterization can then be achieved by letting $p_{\mathcal{H}}(x_{\mathcal{H}}) = p(x_1, x_2)$. Figure 4(d) shows the reparameterization of the PDF shown in Figure 4(a). The normalized cumulative distribution function $C_{\mathcal{H}}(x_{\mathcal{H}})$ is computed and used to sample ξ_u according to the reparameterized PDF $p_{\mathcal{H}}(x_{\mathcal{H}})$. The sampling can be done by drawing a sample ξ in the unit interval and finding the corresponding point ξ_u where $C_{\mathcal{H}}(\xi_u) = \xi$. This is demonstrated in Figure 4(e). Finally, the two-dimensional sample (ξ_1, ξ_2) is obtained by computing $\mathcal{H}(\xi_u)$, resulting in a set of samples as in Figure 4(c). In this example, 64 samples were drawn.

4. Analysis and Discussion

Shirley [Shirley 91] observed that the star discrepancy correlates well with observable errors in computer graphics applications. Figure 5 shows the evolution of several star discrepancy measures of Hilbert sampling against a variable number of strata in two-dimensional space. The star discrepancy was estimated by selecting the maximum star discrepancy of 10 million sampling patterns with N samples.

Vol. 8, No. 3: 41–47

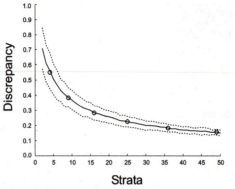

Figure 5. Star discrepancy using the Hilbert curve. The circles show discrepancy of two-dimensional jittered sampling for $2 \times 2, 3 \times 3$, etc. The solid line shows discrepancy for arbitrary numbers of strata using the Hilbert curve. The dotted line shows the variance in the discrepancy estimate at one standard deviation.

Uniform regular (jittered) stratification has n^d strata, where n is the number of subdivisions on each dimension and d is the dimensionality of the domain. Stratification using the Hilbert curve provides a generalization of the jittered stratification to an arbitrary number of strata. When the number of strata is a power of four, the strata configuration obtained by using the Hilbert curve is *exactly* the same as the uniform regular stratification, and consequently has exactly the same discrepancy measure. For other configurations (i.e., $3 \times 3, 5 \times 5$, and so on), the difference between the discrepancy measures of both stratification techniques is negligible. The same behavior is observed for line [Dobkin and Mitchell 93] and triangle [Heinrich 94] discrepancies.

This technique can be extended to adaptive sampling, where one-dimensional intervals are refined recursively according to a given oracle and then mapped to n-dimensional regions. The refinement can be based on neighboring strata (not including the stratum subject to refinement) in order to minimize bias. In general, the space-filling curve approach permits many one-dimensional sampling strategies to be applied to n-dimensional problems.

References

[Bially 69] Theodore Bially. "Space-Filling Curves: Their Generation and Their Application to Bandwidth Reduction." *IEEE Transactions on Information Theory* 15:6 (1969), 658–664.

[Butz 71] Arthur R. Butz. "Alternative Algorithm for Hilbert's Space-Filling Curve." *IEEE Transactions on Computers* 20:4 (1971), 424–426.

[Dobkin and Mitchell 93] David P. Dobkin and Don P. Mitchell. "Random-Edge Discrepancy of Supersampling Patterns." In *Graphics Interface '93*, pp. 62–69. Toronto: Canadian Information Processing Society, 1993.

[Heinrich 94] Stefan Heinrich and Alexander Keller. "Quasi-Monte Carlo Methods in Computer Graphics, Part I: The QMC-Buffer." Technical Report 242/94, University of Kaiserslautern, 1994.

[McCool et al. 01] Michael D. McCool, Chris Wales, and Kevin Moule. "Incremental and Hierarchical Hilbert Order Edge Equation Polygon Rasterization." In *Proceedings of the ACM SIGGRAPH/EUROGRAPHICS Workshop on Graphics Hardware 2001*, pp. 65–72, New York: ACM Press, 2001.

[Sagan 92] Hans Sagan. "On the Geometrization of the Peano Curve and the Arithmetization of the Hilbert Curve." *Internat. J. Mth. Ed. Sc. Tech.* 23 (1992), 403–411.

[Sagan 94] Hans Sagan. *Space-Filling Curves.* New York: Springer-Verlag, 1994.

[Shirley 91] Peter Shirley. "Discrepancy as a Quality Measure for Sample Distributions." In *Proceedings of Eurographics '91*, pp. 183–194, Amsterdam: Elsevier Science Publishers, 1991.

[Voorhies 91] Douglas Voorhies. "Space-Filling Curves and a Measure of Coherence." In *Graphics Gems II*, pp. 26–30. Boston: Academic Press, 1991.

[Warren 03] Henry S. Warren, Jr. *Hacker's Delight.* Boston: Addison-Wesley, 2003.

Web Information:

The source code of Figure 2 as well as the code for sampling an arbitrary PDF is available at http://www.acm.org/jgt/papers/SteiglederMcCool03.

Mauro Steigleder, School of Computer Science, University of Waterloo, 200 University Avenue West, Waterloo, Ontario N2L 3G1, Canada (msteigleder@cgl.uwaterloo.ca)

Michael McCool, School of Computer Science, University of Waterloo, 200 University Avenue West, Waterloo, Ontario N2L 3G1, Canada (mmccool@cgl.uwaterloo.ca)

Received October 25, 2002; accepted in revised form September 23, 2003.

Vol. 8, No. 3: 41–47

Current Contact Information:

Mauro Steigleder, School of Computer Science, University of Waterloo, 200 University Avenue West, Waterloo, Ontario N2L 3G1, Canada (msteigleder@cgl.uwaterloo.ca)

Michael McCool, School of Computer Science, University of Waterloo, 200 University Avenue West, Waterloo, Ontario N2L 3G1, Canada (mmccool@cgl.uwaterloo.ca)

Vol. 5, No. 4: 9–12

Generating Random Points in a Tetrahedron

C. Rocchini and P. Cignoni

Istituto di Elaborazione dell'Informazione – Consiglio Nazionale delle Ricerche

Abstract. This paper proposes a simple and efficient technique to generate uniformly random points in a tetrahedron. The technique generates random points in a cube and folds the cube into the barycentric space of the tetrahedron in a way that preserves uniformity.

1. Introduction

The problem of picking a random point inside a tetrahedron with uniform distribution can arise in various situations, typically in scientific visualization when managing unstructured datasets represented by tetrahedral meshes. In some applications[Cignoni et al. 2000] it is necessary to create more than $O(10^7)$ samples, so an efficient technique to generate a random sample inside a tetrahedron can be very useful.

The method proposed here is a generalization of one of the techniques used by Turk in [Turk 90] to generate a random point inside a triangle. The main idea of his two-dimensional technique is to generate a random point in a parallelogram and reflect it around the center of the parallelogram. In this paper we extend this approach to three dimensions, generating random points with uniform distribution in a cube and folding them into a tetrahedron. Source code of an implementation of this technique is available online.

Vol. 5, No. 4: 9–12

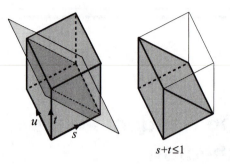

Figure 1. The plane $s + t = 1$ cuts the cube into two equal-volume triangular prisms.

2. Folding a Cube Into a Tetrahedron

Let s, t and u be three numbers chosen from a uniform distribution of random numbers in the interval $[0,1]$ and $\mathbf{a}, \mathbf{b}$ and $\mathbf{c}$ three three-dimensional vectors; the point $s\mathbf{a} + t\mathbf{b} + u\mathbf{c}$ identifies a random point with uniform distribution inside the three-dimensional skewed parallelepiped defined by $\mathbf{a}, \mathbf{b}$ and $\mathbf{c}$. To simplify the discussion we will focus only on s, t and u and how how to *dissect and fold* the cubic parametric space defined by s, t and u into a tetrahedron.

Note that while the area of a triangle is one-half of the area of the corresponding parallelogram, the volume of a tetrahedron is one-sixth of the volume of the corresponding parallelepiped. Therefore, we will dissect and fold it in two and the result again in three parts in order to obtain the $2 \times 3 = 6$ tetrahedra.

The first step is to cut the cube with the plane $s + t = 1$ into two triangular prisms of equal volume. We then fold all the points falling beyond the plane $s + t = 1$ (i.e., in the upper prism) into the lower prism as shown in Figure 1; this can be done by calculating the new (s, t, u) values as shown in Equation 1:

$$(s, t, u) = \begin{cases} (s, t, u) & \text{if } s + t \leq 1, \\ (1 - s, 1 - t, u) & \text{if } s + t > 1. \end{cases} \tag{1}$$

The second step is to cut and fold the resulting triangular prism with the two planes $t + u = 1$ and $s + t + u = 1$. This dissection identifies the three equal volume tetrahedra shown in Figure 2; the folding of the triangular prism into the first tetrahedron can be done by calculating the new (s, t, u) values as shown in Equation 2:

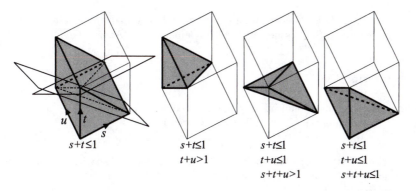

Figure 2. The triangular prism can be cut into three equal volume tetrahedra by the planes $t + u = 1$ and $s + t + u = 1$. Below each tetrahedron we show the inequalities that are satisfied.

$$(s, t, u) = \begin{cases} (s, t, u) & \text{if } s + t + u \leq 1 \\ \begin{cases} (s, 1 - u, 1 - s - t) & \text{if } t + u > 1 \\ (1 - t - u, t, s + t + u - 1) & \text{if } t + u \leq 1 \end{cases} & \text{if } s + t + u > 1 \end{cases}$$
(2)

3. Alternative Approach

A technique to generate a random point inside a tetrahedron is described in [Turk 90]: three random numbers are chosen from a uniform distribution; the cube root of the first number is used to pick a triangle that is parallel to the base of the tetrahedron, and then the two remaining numbers are used to pick a random point on that triangle. This technique is approximately twice as slow as the one proposed here. On the other hand, it allows stratification. Another of the advantages of our technique is that the barycentric coordinates of the sampled point are available at the end of the process. Therefore, it is easy to find interpolate values for attributes of the sampled point that were originally defined onto the tetrahedron vertices.

References

[Cignoni et al. 2000] P. Cignoni, C. Rocchini, and R. Scopigno. "Metro 3d: Measuring error on simplified tetrahedral complexes." Technical Report B4-09-00, I.E.I. – C.N.R., Pisa, Italy, Feb. 2000.

[Turk 90] Greg Turk. "Generating random points in triangles." In *Graphics Gems*, edited by A. S. Glassner, pp. 24–28. Cambridge, MA: Academic Press, 1990.

Web Information:

Source code and additional information are available online at http://www.acm.org/jgt/papers/RocchiniCignoni00

C. Rocchini, Istituto di Elaborazione dell'Informazione – Consiglio Nazionale delle Ricerche, CNR Research Park, Via V. Alfieri 1, 56010 S.Giuliano (Pisa), Italy (rocchini@iei.pi.cnr.it)

P. Cignoni, Istituto di Elaborazione dell'Informazione – Consiglio Nazionale delle Ricerche, CNR Research Park, Via V. Alfieri 1, 56010 S.Giuliano (Pisa), Italy (cignoni@iei.pi.cnr.it)

Received May 31, 2000; accepted in revised form October 17, 2000.

Updated Web Information:

http://vcg.isti.cnr.it/jgt/tetra.htm

Current Contact Information:

C. Rocchini, Istituto Geografico Militare, Via Cesare Battisti 10, Firenze, Italy (ad2prod@geomil.esercito.difesa.it)

P. Cignoni, ISTI-CNR, Via Moruzzi, 1, 56124 Pisa, Italy (p.cignoni@isti.cnr.it)

Vol. 4, No. 4: 11–22

An RGB-to-Spectrum Conversion
for Reflectances

Brian Smits
University of Utah

Abstract. This paper presents a solution to the problem of using RGB data created by most modeling systems in a spectrally-based renderer. It differs from previous methods in that it attempts to create physically plausible spectra for reflectances. The method searches the metamer space for a spectrum that best fits a set of criteria. The results are used in an algorithm that is simple and efficient enough to be used within the rendering process for both importing models and for texture mapping.

1. Introduction

The desire for accuracy and realism in images requires a physically-based rendering system. Often this can mean using a full spectral representation, as RGB representations have limitations in some situations [Hall 89]. The spectral representation does come at some cost, not the least of which is that most available models and model formats specify materials and lights in terms of RGB triples. Using these models conveniently requires the conversion of RGB values into spectra [Glassner 89]. The conversion process should result in spectra that are suitable for use in a physically-based rendering system and be efficient, both for data import and for texturing operations.

The main difficulty in converting from RGB to spectra is that for any positive RGB color there is an infinite number of spectra that can produce that color. These spectra are known as *metamers* of each other. Although all these spectra produce the same color, when they are used as reflectances and illuminated by nonconstant spectra the resulting spectra are not usually metamers.

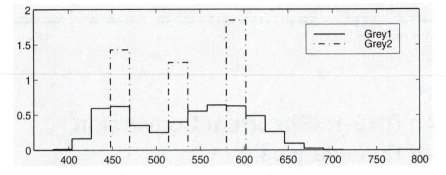

Figure 1. Two spectral representations for grey, G_1 and G_2.

There is not enough information to determine which of the metamers for a particular color is the "right" metamer. Instead, the goal is to pick a "good" metamer.

In the physical world, spectra are functions defined over some continuous range of wavelengths. This is an impractical representation for a rendering algorithm, so a finite dimensional representation must be chosen. These range from piecewise constant basis functions to optimally chosen point samples [Meyer 87] to optimal basis functions for a specific scene [Peercy 93]. This paper uses piecewise constant spectra, as the relatively narrow and nonoverlapping support of the basis functions make them efficient to use and capable of representing fairly saturated colors. Since they cover every wavelength in the visible spectrum, the filtering issues raised by point samples are handled automatically. For the rest of this paper, a spectrum will refer to a vector of weights for a set of piecewise constant basis functions.

The linear transform from a spectrum to the RGB color space is defined by the standard XYZ matching functions and the matrix converting from XYZ to RGB as described in many sources (e.g., [Rogers 85], [Wyszecki, Stiles 82]). Note that the XYZ matching functions are defined from 360 nm to 800 nm, while in practice rendering systems usually use a significantly smaller interval in order to focus on the region where the eye is more sensitive.

Most of the colors that will need to be transformed while reading in a model, or while doing texture mapping, are reflectances. Reflectances should all be between zero and one in order to satisfy physical laws. Reflectances of natural materials also tend to be smooth [Maloney 86]. Additionally, spectra with a minimal amount of variation are better for computational reasons. This can be seen by looking at two possible representations for 50% grey in Figure 1. Although both are the same color, they perform very differently in a rendering system. The results of multiplying the two curves is shown in Figure 2.

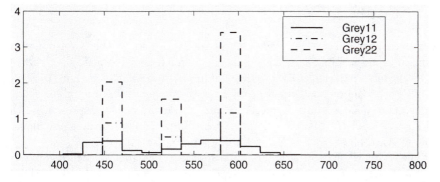

Figure 2. The three products, G_1G_1, G_1G_2, and G_2G_2.

None of the resulting curves is grey or average 25%. A global illumination algorithm using Grey2 would diverge very quickly, while a rendering system using Grey1, or a combination, would produce unintuitive results and visually disturbing color shifts. Both these cases are possible in the real world, however, they don't match our intuition of a grey surface illuminated by a white light resulting in a grey appearance. Minimal variation over the spectrum, with peaks and troughs being as wide and as close to average as possible, reduce these problems.

Previous approaches to converting RGB colors to spectra have not focused on creating spectra that are suitable for reflectances. One approach [Glassner 95] takes the spectra for each of the three phosphors and weights them by the RGB triple. This obviously works. However, the spectra for the phosphors are quite irregular resulting in reflectances that are significantly greater than 1, and a white RGB results in a spectrum that is far from constant. Other approaches by Glassner [Glassner 89] and Sun et al. [Sun et al. 99] choose three smooth functions and use linear combinations of those functions based on the RGB values. These approaches can result in spectra with negative regions. A last and simplest approach was used by Shirley and Marschner [Shirley 98]. This approach simply uses the blue value for the first $n/3$ coefficients, the green value for the next $n/3$, and the red for the last $n/3$ coefficients. The number of basis functions were constrained to be a multiple of three. This method does not produce a spectrum that converts to the right color, but it can be fairly close, and results in spectra that are guaranteed to be valid reflectances. The method is tailored to spectra defined over the interval [400, 700] and does not work satisfactorily for intervals significantly different from this. Also, the resulting curves did not look very plausible. Ideally, a good conversion for reflectances would have the bounded properties of the Shirley and Marschner approach and the smoothness and exact match properties of the method described by Glassner.

2. Algorithm

Given a spectrum $s \in \mathbb{R}^n$ and a standard RGB color c, the XYZ matching functions and an XYZ-to-RGB transform define a transformation matrix $A \in \mathbb{R}^{3 \times n}$ from spectra to RGB. We are looking for a good s such that $As = c$.

This is an underconstrained problem as there is an infinite number of solutions. Let s_0 be a solution and $N = \text{null}(A)$ be the matrix representing the null space of A ($N \in \mathbb{R}^{n \times n-3}$, and every column of N is orthogonal to each row of A), then $s^x = s^0 + Nx$ is a valid solution for all $x \in \mathbb{R}^{n-3}$.

The set of s^x are all metamers of each other. The goal is to choose a good metamer which can be defined as $\forall i \; s_i^x \geq 0$ and s^x is smooth. Smoothness can be defined in many ways. After some experimentation, it seemed that the best solution was to look at the differences between adjacent basis functions as a measure of smoothness. This can be done by minimizing the two-norm of the difference vector $d \in \mathbb{R}^{n-1} : d_i = s_i - s_{i+1}$, a standard approach in vision when using snakes to fit a curve [Trucco, Verri 98]. Constant regions are considered ideal, and if the function needs to vary, several small steps are better than a single large step. This tends to create wide peaks and valleys, and works to minimize the overall difference. This fits the definition of "good" as described in Section 1.

In order to make these spectra valid reflectances, the values are constrained to be greater than zero. Ideally, a constraint keeping the spectrum less than one would also be added but, as will be explained later, this can be impossible to satisfy for some representations of the visible spectrum. Instead, a penalty factor is added to the cost function when any value of the spectrum is greater than one.

Although all of the previous steps can be done very simply and efficiently in Matlab [Pärt-Enander et al. 96], they are still too costly to do for every conversion. The immediate solution to this is to exploit the linearity of the conversion and create spectra for red, green, and blue. The plot of these is shown in Figure 3.

The immediate problem with this approach can be seen by looking at the sum. Its color is white, and it should be as smooth as possible. As shown in Figure 4, the conversion for white looks much better than the sum of the three primary colors. A reasonable solution is to continue to exploit the linearity of the conversions, and create spectra for red, green, blue, cyan, magenta, yellow, and white.

The spectra for cyan, magenta, and yellow are shown in Figure 5. In the same way that the white spectrum is better than a sum of red, green, and blue, the cyan, magenta, and yellow spectra are better than the sum of the appropriate red, green, and blue spectra.

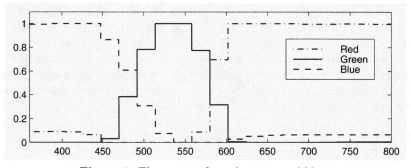

Figure 3. The spectra for red, green, and blue.

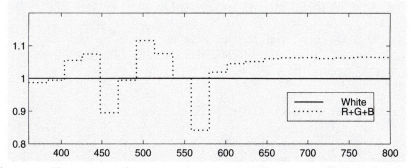

Figure 4. The spectrum for white versus the sum of the spectra for red, green, and blue.

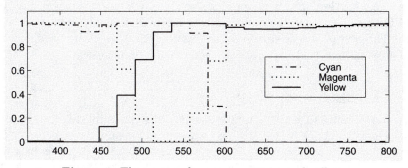

Figure 5. The spectra for cyan, magenta, and yellow.

Ideally, the conversion would use as much of the constant white spectrum as possible, followed by the secondary colors, and mix in primary colors only as needed. Any color can be expressed as a sum of white, plus one of cyan, magenta, or yellow, plus one of red, green, or blue. The conversion process works by subtracting as much of the wider spectra (first white, then either cyan, magenta, or yellow) as possible before converting the remainder using the red, green, or blue spectra. This can be expressed in pseudocode as follows for a spectrum where red is less than green and blue:

```
Spectrum RGBToSpectrum(red,green,blue)
    Spectrum ret = 0;
    if(red ≤green && red ≤ blue)
        ret += red * whiteSpectrum;
        if(green ≤ blue)
            ret += (green - red) * cyanSpectrum;
            ret += (blue - green)* blueSpectrum;
        else
            ret += (blue - red) * cyanSpectrum;
            ret += (green - blue)* greenSpectrum;
    else if(green ≤ red && green ≤ blue)
        ⋮
    else // blue ≤ red && blue ≤ green
        ⋮
```

The other two cases are similar.

3.   Unresolved Issues

The algorithm described in the previous section gives "good" conversions from an RGB triple to a spectrum, but unfortunately that is only part of the solution. This section discusses some of the issues not addressed by the algorithm and gives options that may partially address the problems.

3.1.   Convergence Problems

It is often possible to get into situations where it is impossible to create spectra between zero and one for the various components (usually the red curve). For spectra defined from 400 nm to 700 nm and an XYZ-to-RGB matrix based on Sony monitor data with a white point set at $(0.333, 0.333)$[1], it is not possible

---

[1] $R_{xy} = (0.625, 0.34), G_{xy} = (0.28, 0.595), B_{xy} = (0.155, 0.07), W_{xy} = (0.333, 0.333)$

to get the red curve below one. The area of the spectrum contributing solely to red is not large enough. Even though all the matching functions are low past 700 nm, the red is still significant and more importantly, the other two matching functions are negative.

Two factors are responsible for causing curves to rise above one: the width of the visible spectrum and the chromaticity coordinates for the primaries used to create the XYZ-to-RGB matrix. The width of the spectrum can be increased. However, for small numbers of spectral samples (nine or less), uniform bin size results in poor sampling of the regions of the spectrum where the matching functions are changing more frequently. As the primaries get more saturated, this problem gets worse.

Consider three monochromatic primaries. The only way to (almost) match these is to have a zero for every basis function except the one containing that wavelength. For the sum of the red, green, and blue curves to have the same intensity as a constant unit spectrum, the value of the single nonzero coefficient must be much larger than one.

3.2. White Point Considerations

The white point of a monitor is the color produced by an RGB triple of (1, 1, 1), and can vary significantly between different types of monitors. The whiteness constancy effect causes us to perceive these different colors as white, as long as they can't be easily compared with each other, such as when two different monitors are placed next to each other. Because of this effect, the conversion to and from spectra should be set up with a white point of (0.333, 0.333) for most uses. This is the only way white can be a constant spectrum.

There are three other options that may make sense in various circumstances. First, if spectral data is mixed with RGB data it may be more reasonable to convert to spectra using a white point of (0.333, 0.333) and convert from spectra using the correct white point. The RGB data won't match exactly, but should be reasonably close, and the reflectances will be valid.

If achieving an accurate match is important and spectral data will also be used, the correct white point can be used for conversions in both directions. Spectra may no longer be automatically bounded by one due to white no longer being constant, a problem discussed in the next section. Although using the same white point makes the conversion invertible, since white is no longer constant, white light shining on a white object will exhibit color shifts. A variation on this option is to use the correct white point for converting materials and using a white point of (0.333, 0.333) for lights. This will allow matching the RGB values of a conventional renderer for white lights and getting very close for others.

3.3. Scale Factor for Reflectances

The algorithm described above ignores scaling issues in the conversion from spectra to RGB, which is a function of the area under the XYZ matching functions (roughly 106), conversion to lumens, and the white point luminance chosen in the XYZ to RGB transform. This means that a grey RGB triple (0.7, 0.7, 0.7) may be converted into a constant spectrum with values nowhere near 0.7. This is clearly not going to work correctly.

This is easily solved by making the white point luminance of XYZ to RGB transform exactly equal to the area under the XYZ matching functions weighted by the lumens conversion factor. This means a constant spectrum with a value of 1 converts to (1, 1, 1), and therefore the inverse will have the right range. If this is not done, or if chromaticity coordinates or white points must be set so that the various curves have values greater than one, it may be necessary to scale or clip the resulting spectrum if values between zero and one are desired. This will introduce some error for those spectra.

Another issue has to do with specifying appearance rather than reflectance. Most users of modeling software set white walls to a reflectance of 1, whereas white paint actually has a reflectance that is usually below 0.8. This is because they want a white wall, not a grey wall and 0.8 definitely looks grey. This can cause significant problems in global illumination systems as solutions will tend not to converge. One solution that works in practice is to scale all incoming data by a factor of about 0.8 (roughly the reflectance of ANSI safety white [Rea 93]).

3.4. Gamma Correction

Until this point, gamma correction has been ignored. In many ways it is a separate and complicated problem all its own [Poynton 96]. The spectral conversion process should be done in the physical intensity domain since it relies on linearity properties that are no longer valid after the gamma correction used to get display values. If gamma is correctly treated as a display issue, there are no problems.

More commonly, however, the modeling package will not do any (or enough) gamma correction, and the spectral renderer will do gamma correction before displaying the result. In this case, the RGB data in the model are display values, and an inverse gamma correction should be done before the spectral conversion process in order to get physical intensities.

Figure 6. Rendering of direct illumination using RGB renderer (left) and spectral renderer (right). (See Color Plate III.)

4. Examples and Discussion

The algorithm described in Section 2 was tested on a simple environment with many highly saturated objects. The three images on the walls are texture maps of the three bright faces of the RGB color cube (R, G, or B = 1). The lights used for the first test (Figure 6) are white. The white point for the RGB conversion was set at (0.333, 0.333) as described in Section 3. The exact data for the conversion are given at the end of this paper. The left image in Figure 6 was rendered with the ray tracer inside Maya [Alias|Wavefront 98], while the right image was rendered with a spectrally- and physically-based renderer. In both cases, only direct illumination was used. The spectral renderer computed the texture maps in RGB space and converted to a spectrum when the value was needed by the shader. The two pictures are not exact but are very close, and the colors are indistinguishable visually between the two images. Because the light sources are white, and only direct illumination is being rendered, the mathematics say the values should match exactly. Differences are likely due to variations in BRDF models and lighting calculations between the two renderers, as well as to the final scaling done to convert to displayable intensities. As the goal of this paper is to be able to bring RGB data into a spectral renderer while preserving appearance, the minor differences are not significant.

The second set of images (shown in Figure 7) was created with lights that are not white. In addition, indirect illumination was computed by the spectrally-based renderer (using standard Monte Carlo methods). Although there is now a visual difference, these differences are unavoidable. As discussed in Section 1, once nonconstant spectra are multiplied there is no right answer. The differences are not large enough to hinder color selection in standard RGB modeling/rendering systems, or to create difficulties in importing models with RGB data.

Figure 7. Rendering with nonwhite light sources using RGB renderer with direct lighting (left) and spectral render with direct and indirect illumination (right). (See Color Plate IV.)

The conversion method presented here is guaranteed to be invertible and is efficient enough to be used within the texture stage of the renderer. It produces plausible reflectances that capture the appearance of the surfaces even under nonwhite illumination and with multiple interreflections. Although this paper uses uniformly-sized bins to represent the spectra, that fact was not used by the algorithm. The only critical restriction on the representation is that a matrix can be used to convert the spectral representation to RGB values.

It is possible to find situations where the results do not match reasonably well, such as highly saturated lights or transparent filters with (relatively) selective transparency. However, that is almost certainly unavoidable with any method. While the resulting spectra may look plausible and have nice properties within global illumination algorithms, there is no guarantee they are representative of any real material. They simply allow the same renderer to render models with RGB data, spectral data, or both.

Acknowledgments. I would like to thank Peter Shirley for helpful discussions during the writing of this paper. Thanks to Peter Shirley, Amy Gooch, Simon Premoze, and the reviewers for their comments. Also, thanks to Alias|Wavefront for providing Maya. This work was supported by NSF awards 97-31859 and 97-20192.

Data. The data presented here is based on a 10 bin spectral representation from 380 nm to 720 nm. The bins are all equal size. The data for the XYZ to RGB conversion are $R_{xy} = (0.64, 0.33)$, $G_{xy} = (0.3, 0.6)$, $B_{xy} = (0.15, 0.06)$, $W_{xy} = (0.333, 0.333)$, $W_y = 106.8$. The phosphor chromaticities are based on the IUT-R BT.709 standard as discussed by a W3C recommendation describing a default color space for the Internet [Stokes et al. 96]. The Matlab scripts used to generate the data may be found at the web site listed at the end of the article.

bin	white	cyan	magenta	yellow	red	green	blue
1	1.0000	0.9710	1.0000	0.0001	0.1012	0.0000	1.0000
2	1.0000	0.9426	1.0000	0.0000	0.0515	0.0000	1.0000
3	0.9999	1.0007	0.9685	0.1088	0.0000	0.0273	0.8916
4	0.9993	1.0007	0.2229	0.6651	0.0000	0.7937	0.3323
5	0.9992	1.0007	0.0000	1.0000	0.0000	1.0000	0.0000
6	0.9998	1.0007	0.0458	1.0000	0.0000	0.9418	0.0000
7	1.0000	0.1564	0.8369	0.9996	0.8325	0.1719	0.0003
8	1.0000	0.0000	1.0000	0.9586	1.0149	0.0000	0.0369
9	1.0000	0.0000	1.0000	0.9685	1.0149	0.0000	0.0483
10	1.0000	0.0000	0.9959	0.9840	1.0149	0.0025	0.0496

References

[Alias|Wavefront 98] Alias|Wavefront. *Maya v. 1.5.* Toronto: Alias|Wavefront, 1998.

[Glassner 89] A. S. Glassner. "How to Derive a Spectrum from an RGB Triplet." *IEEE Computer Graphics and Applications.* 9(4):95–99 (July 1989).

[Glassner 95] A. S. Glassner. *Principles of Digital Image Synthesis.* San Francisco: Morgan-Kaufman, 1995.

[Hall 89] R. Hall. *Illumination and Color in Computer Generated Imagery.* New York: Springer-Verlag, 1989

[Maloney 86] L. T. Maloney. "Evaluation of Linear Models of Surface Spectral Reflectance with Small Number of Parameters." *Journal of the Optical Society of America* 3(10): 1673–1683, 1986.

[Meyer 87] G. W. Meyer. "Wavelength Selection for Synthetic Image Generation." Technical Report CIS-TR-87-14, University of Oregon (November 1987).

[Pärt-Enander et al. 96] E. Pärt-Enander, A. Sjöberg, B. Melin, and R. Isaksson. *The Matlab Handbook.* Harlow, England: Addison-Wesley, 1996

[Peercy 93] M. S. Peercy. "Linear Color Representations for Full Spectral Rendering." In *Proceedings of SIGGRAPH 93, Computer Graphics Proceedings, Annual Conference Series,* edited by James T. Kajiya, pp. 191–198, New York: ACM Press, 1993.

[Poynton 96] C. Poynton. *A Technical Introduction to Digital Video.* New York: Wiley and Sons, 1996.
http://www.inforamp.net/poynton/Poynton-T-I-Digital-Video.html

[Rea 93] M. S. Rea, editor. *The Illumination Engineering Society Lighting Handbook,* 8th edition, New York: Illumination Engineering Society, 1993.

[Rogers 85] D. F. Rogers. *Procedural Elements for Computer Graphics.* New York: McGraw-Hill, 1985.

[Shirley 98] P. Shirley. Personal Correspondance, 1998.

[Stokes et al. 96] M. Stokes, M. Anderson, S. Chandrasekar, and R. Motta. "A Standard Default Color Space for the Internet—sRGB." 1996.
http://www.w3.org/Graphics/Color/sRGB

[Sun et al. 99] T. Sun, F. D. Fracchia, T. W. Calveret, and M. S. Drew. "Deriving Spectra from Colors and Rendering Light Interference." *IEEE Computer Graphics and Applications* 19(4):61–67 (July/August 1999)

[Trucco, Verri 98] E. Trucco and A. Verri. *Introductory Techniques for 3-D Computer Vision.* Englewood Cliffs, NJ: Prentice Hall, 1998.

[Wyszecki, Stiles 82] G. Wyszecki and W. S. Stiles. *Color Science: Concepts and Methods, Quantitative Data and Formulae.* New York: Wiley, 1982.

Web Information:

http://www.acm.org/jgt/papers/Smits99

Brian Smits, University of Utah, Department of Computer Science, 50 S. Central Campus Drive, Room 3190, Salt Lake City, UT 84112-9205 (bes@cs.utah.edu)

Received August 31, 1999; accepted December 20, 1999.

Current Contact Information:

Brian Smits, Pixar Animation Studios, 1200 Park Ave., Emeryville, CA 94608 (bes@pixar.com)

Vol. 5, No. 2: 25–32

An Anisotropic Phong BRDF Model

Michael Ashikhmin and Peter Shirley

University of Utah

Abstract. We present a BRDF model that combines several advantages of the various empirical models currently in use. In particular, it has intuitive parameters, is anisotropic, conserves energy, is reciprocal, has an appropriate non-Lambertian diffuse term, and is well-suited for use in Monte Carlo renderers.

1. Introduction

Physically-based rendering systems describe reflection behavior using the *bidirectional reflectance distribution function* (BRDF). For a detailed discussion of the BRDF and its use in computer graphics see the volumes by Glassner [Glassner 95]. At a given point on a surface the BRDF is a function of two directions, one toward the light and one toward the viewer. The characteristics of the BRDF will determine what "type" of material the viewer thinks the displayed object is composed of, so the choice of BRDF model and its parameters is important. We present a new BRDF model that is motivated by practical issues. A full rationalization for the model and comparison with previous models is provided in a separate technical report [Ashikhmin, Shirley 00].

The BRDF model described in this paper is inspired by the models of Ward [Ward 92], Schlick [Schlick 94], and Neumann et al. [Neumann et al. 99]. However, it has several desirable properties not previously captured by a single model. In particular, it

1. obeys energy conservation and reciprocity laws,

2. allows *anisotropic* reflection, giving the streaky appearance seen on brushed metals,

3. is controlled by intuitive parameters,

4. accounts for *Fresnel behavior*, where specularity increases as the incident angle goes down,

5. has a nonconstant diffuse term, so the diffuse component decreases as the incident angle goes down,

6. is well-suited to Monte Carlo methods.

The model is a classical sum of a "diffuse" term and a "specular" term.

$$\rho(\mathbf{k}_1, \mathbf{k}_2) = \rho_s(\mathbf{k}_1, \mathbf{k}_2) + \rho_d(\mathbf{k}_1, \mathbf{k}_2). \tag{1}$$

For metals, the diffuse component ρ_d is set to zero. For "polished" surfaces, such as smooth plastics, there is both a diffuse and specular appearance and neither term is zero. For purely diffuse surfaces, either the traditional Lambertian (constant) BRDF can be used, or the new model with low specular exponents can be used for slightly more visual realism near grazing angles. The model is controlled by four parameters:

- R_s: a color (spectrum or RGB) that specifies the specular reflectance at normal incidence.

- R_d: a color (spectrum or RGB) that specifies the diffuse reflectance of the "substrate" under the specular coating.

- n_u, n_v: two Phong-like exponents that control the shape of the specular lobe.

We now illustrate the use of the model in several figures. The remainder of the paper deals with specifying and implementing the model. Figure 1 shows spheres with $R_d = 0$ and varying n_u and n_v. The spheres along the diagonal have $n_u = n_v$ so have a look similar to the traditional Phong model. Figure 2 shows another metallic object. This appearance is achieved by using the "right" mapping of tangent vectors on the surface. Figure 3 shows a "polished" surface with $R_s = 0.05$. This means the diffuse component will dominate for near-normal viewing angles. However, as the viewing angle becomes oblique the specular component dominates despite its low near-normal value. Figure 4 shows the model for a diffuse surface. Note how the ball on the right has highlights near the edge of the model, and how the constant-BRDF ball on the left is more "flat." The highlights produced by the new model are present in the measured BRDFs of some paints, so are desirable for some applications [Lafortune et al. 97].

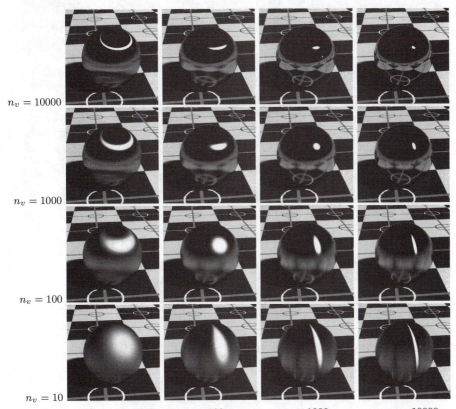

$n_v = 10000$

$n_v = 1000$

$n_v = 100$

$n_v = 10$

$n_u = 10 \qquad n_u = 100 \qquad n_u = 1000 \qquad n_u = 10000$

Figure 1. Metallic spheres for various exponents. (See Color Plate XVIII.)

Figure 2. A cylinder with the appearance of brushed metal created with the new model with $R_d = 0, R_s = 0.9, n_u = 10, n_v = 100$.

Figure 3. Three views for $n_u = n_v = 400$ and a red substrate. (See Color Plate XVII.)

Figure 4. An image with a Lambertian sphere (left) and a sphere with $n_u = n_v = 5$. After a figure from Lafortune et al. [Lafortune et al. 97]. (See Color Plate XIX.)

2.   The Model

The specular component ρ_s of the BRDF is:

$$\rho_s(\mathbf{k}_1, \mathbf{k}_2) = \frac{\sqrt{(n_u+1)(n_v+1)}}{8\pi} \frac{(\mathbf{n} \cdot \mathbf{h})^{n_u \cos^2 \phi + n_v \sin^2 \phi}}{(\mathbf{h} \cdot \mathbf{k})\max((\mathbf{n} \cdot \mathbf{k}_1), (\mathbf{n} \cdot \mathbf{k}_2))} F((\mathbf{k} \cdot \mathbf{h})). \quad (2)$$

In our implementation we use Schlick's approximation to the Fresnel fraction [Schlick 94]:

$$F((\mathbf{k} \cdot \mathbf{h})) = R_s + (1 - R_s)(1 - (\mathbf{k} \cdot \mathbf{h}))^5, \quad (3)$$

$\mathbf{a} \cdot \mathbf{b}$	scalar (dot) product of vectors $\mathbf{a}$ and $\mathbf{b}$
$\mathbf{k}_1$	normalized vector to light
$\mathbf{k}_2$	normalized vector to viewer
$\mathbf{n}$	surface normal
$\mathbf{u}, \mathbf{v}$	tangent vectors that form an orthonormal basis along with $\mathbf{n}$.
$\rho(\mathbf{k}_1, \mathbf{k}_2)$	BRDF
$\mathbf{h}$	normalized half-vector between $\mathbf{k}_1$ and $\mathbf{k}_2$
$p(\mathbf{k})$	probability density function for reflection sampling rays
$F(\cos \theta)$	Fresnel reflectance for incident angle θ

Table 1. Important terms used in the paper.

where R_s is the material's reflectance for the normal incidence. It is not necessary to use trigonometric functions to compute the exponent in Equation 2, so the specular BRDF can be written:

$$\rho_s(\mathbf{k}_1, \mathbf{k}_2) = \frac{\sqrt{(n_u + 1)(n_v + 1)}}{8\pi} \frac{(\mathbf{n} \cdot \mathbf{h})^{\frac{(n_u (\mathbf{h} \cdot \mathbf{u})^2 + n_v (\mathbf{h} \cdot \mathbf{v})^2)}{(1 - (\mathbf{h} \cdot \mathbf{n})^2)}}}{(\mathbf{h} \cdot \mathbf{k}) \max((\mathbf{n} \cdot \mathbf{k}_1), (\mathbf{n} \cdot \mathbf{k}_2))} F((\mathbf{k} \cdot \mathbf{h})). \quad (4)$$

It is possible to use a Lambertian BRDF together with our specular term in a way similar to that which is done for most models [Schlick 94], [Ward 92]. However, we use a simple angle-dependent form of the diffuse component which accounts for the fact that the amount of energy available for diffuse scattering varies due to the dependence of the specular term's total reflectance on the incident angle. In particular, diffuse color of a surface disappears near the grazing angle because the total specular reflectance is close to one in this case. This well-known effect cannot be reproduced with a Lambertian diffuse term and is therefore missed by most reflection models. Another, perhaps

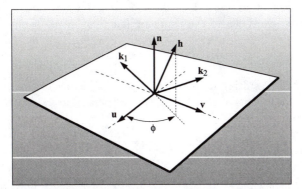

Figure 5. Geometry of reflection. Note that $\mathbf{k}_1$, $\mathbf{k}_2$, and $\mathbf{h}$ share a plane, which usually does not include $\mathbf{n}$.

more important, limitation of the Lambertian diffuse term is that it must be set to zero to ensure energy conservation in the presence of a Fresnel-weighted term.

Our diffuse term is:

$$\rho_d(\mathbf{k}_1, \mathbf{k}_2) = \frac{28 R_d}{23\pi}(1 - R_s)\left(1 - \left(1 - \frac{(\mathbf{n} \cdot \mathbf{k}_1)}{2}\right)^5\right)\left(1 - \left(1 - \frac{(\mathbf{n} \cdot \mathbf{k}_2)}{2}\right)^5\right).$$

(5)

Note that our diffuse BRDF does not depend on n_u and n_v. The somewhat unusual leading constant is designed to ensure energy conservation.

3.   Using the BRDF Model in a Monte Carlo Setting

In a Monte Carlo setting we are interested in the following problem: Given $\mathbf{k}_1$, generate samples of $\mathbf{k}_2$ with a distribution whose shape is similar to the cosine-weighted BRDF. This distribution should be a *probability density function* (pdf), and we should be able to evaluate it for a given randomly-generated $\mathbf{k}_2$. The key part of our thinking on this is inspired by discussions by Zimmerman [Zimmerman 98] and by Lafortune [Lafortune, Willems 94] who point out that greatly undersampling a large value of the integrand is a serious error while greatly oversampling a small value is acceptable in practice. The reader can verify that the densities suggested below have this property.

We first generate a half-vector $\mathbf{h}$ using the following pdf:

$$p_h(\mathbf{h}) = \frac{\sqrt{(n_u + 1)(n_v + 1)}}{2\pi}(\mathbf{n} \cdot \mathbf{h})^{n_u \cos^2 \phi + n_v \sin^2 \phi}.$$

(6)

To evaluate the rendering equation we need both a reflected vector $\mathbf{k}_2$ and a probability density function $p(\mathbf{k}_2)$. Note that if you generate $\mathbf{h}$ according to $p_h(\mathbf{h})$ and then transform to the resulting $\mathbf{k}_2$:

$$\mathbf{k}_2 = -\mathbf{k}_1 + 2(\mathbf{k}_1 \cdot \mathbf{h})\mathbf{h},$$

(7)

the density of the resulting $\mathbf{k}_2$ is **not** $p_h(\mathbf{k}_2)$. This is because of the difference in measures in $\mathbf{h}$ and $\mathbf{v}_2$ space. So the actual density $p(\mathbf{k}_2)$ is

$$p(\mathbf{k}_2) = \frac{p_h(\mathbf{h})}{4(\mathbf{k}_1 \cdot \mathbf{h})}.$$

(8)

Monte Carlo renderers that use this method converge reasonably quickly (Figure 6). In an implementation where the BRDF is known to be this model, the estimate of the rendering equation is quite simple as many terms cancel out.

Figure 6. A closeup of the model implemented in a path tracer with 9, 26, and 100 samples. (See Color Plate XX.)

Note that it is possible to generate an **h** vector whose corresponding vector $\mathbf{k}_2$ will point inside the surface, i.e., $(\mathbf{k}_2 \cdot \mathbf{n}) < 0$. The weight of such a sample should be set to zero. This situation corresponds to the specular lobe going below the horizon and is the main source of energy loss in the model. This problem becomes progressively less severe as n_u and n_v become larger.

The only thing left now is to describe how to generate **h** vectors with pdf of Equation 7. We start by generating **h** with its spherical angles in the range $(\theta, \phi) \in [0, \frac{\pi}{2}] \times [0, \frac{\pi}{2}]$. Note that this is only the first quadrant of the hemisphere. Given two random numbers (ξ_1, ξ_2) uniformly distributed in $[0, 1]$, we can choose

$$\phi = \arctan\left(\sqrt{\frac{n_u + 1}{n_v + 1}} \tan\left(\frac{\pi \xi_1}{2}\right)\right), \tag{9}$$

and then use this value of ϕ to obtain θ according to

$$\cos \theta = (1 - \xi_2)^{\frac{1}{n_u \cos^2 \phi + n_v \sin^2 \phi + 1}}. \tag{10}$$

To sample the entire hemisphere, the standard manipulation where ξ_1 is mapped to one of four possible functions depending on whether it is in $[0, 0.25)$, $[0.25, 0.5)$, $[0.5, 0.75)$, or $[0.75, 1.0)$ is used. For example for $\xi_1 \in [0.25, 0.5)$, find $\phi(1 - 4(0.5 - \xi_1))$ via Equation 9, and then "flip" it about the $\phi = \pi/2$ axis. This ensures full coverage and stratification.

For the diffuse term it would be possible to do an importance sample with a density close to the cosine-weighted BRDF (Equation 5) in a way similar to that described by Shirley et al. [Shirley et al. 97], but we use a simpler approach and generate samples according to the cosine distribution. This is sufficiently close to the complete diffuse BRDF to substantially reduce variance of the Monte Carlo estimation. To generate samples for the entire BRDF, simply use a weighted average of 1 and $p(\mathbf{k}_2)$.

References

[Ashikhmin, Shirley 00] Michael Ashikhmin and Peter Shirley. "An anisotropic phong light reflection model." Technical Report UUCS-00-014, Computer Science Department, University of Utah, June 2000.

[Glassner 95] Andrew S. Glassner. *Principles of Digital Image Synthesis.* San Francisco: Morgan-Kaufman, 1995.

[Lafortune, Willems 94] Eric P. Lafortune and Yves D. Willems. "Using the modified phong BRDF for physically based rendering." Technical Report CW197, Computer Science Department, K.U.Leuven, November 1994.

[Lafortune et al. 97] Eric P. F. Lafortune, Sing-Choong Foo, Kenneth E. Torrance, and Donald P. Greenberg. "Non-linear approximation of reflectance functions." In *Proceedings of SIGGRAPH 97, Computer Graphics Proceedings, Annual Conference Series*, edited by Turner Whitted, pp. 117–126, Reading, MA: Addison Wesley, 1997.

[Neumann et al. 99] László Neumann, Attila Neumann, and László Szirmay-Kalos. "Compact metallic reflectance models." *Computer Graphics Forum*, 18(13): 161–172 (1999).

[Schlick 94] Christophe Schlick. "An inexpensive BRDF model for physically-based rendering." *Computer Graphics Forum*, 13(3): 233–246 (1994).

[Shirley et al. 97] Peter Shirley, Helen Hu, Brian Smits, and Eric Lafortune. "A practitioners' assessment of light reflection models." In *The Fifth Pacific Conference on Computer Graphics and Applications : October 13-16, 1997 Seoul National University, Seoul, Korea : Proceedings*, pp. 40–49, Los Alamitos, CA: IEEE Computer Society, 1997.

[Ward 92] Gregory J. Ward. "Measuring and modeling anisotropic reflection." *Computer Graphics (Proc. SIGGRAPH '92)*, 26(4): 265–272 (July 1992).

[Zimmerman 98] Kurt Zimmerman. *Density Prediction for Importance Sampling in Realistic Image Synthesis.* PhD thesis, Indiana University, June 1998.

Web Information:

http://www.acm.org/jgt/papers/AshikhminShirley00

Michael Ashikhmin, University of Utah, Computer Science Department, 50 S. Central Campus Drive, Salt Lake City, UT 84112 (michael@cs.utah.edu)

Peter Shirley, University of Utah, Computer Science Department, 50 S. Central Campus Drive, Salt Lake City, UT 84112 (shirley@cs.utah.edu)

Received July 10, 2000; accepted August 11, 2000.

Current Contact Information:

Michael Ashikhmin, SUNY at Stony Brook, Computer Science Department, 1434 Computer Science Building, State University of New York, Stony Brook, NY 11794 (ash@cs.sunysb.edu)

Peter Shirley, University of Utah, Computer Science Department, 50 S. Central Campus Drive, Salt Lake City, UT 84112 (shirley@cs.utah.edu)

Part VII

Image Processing

Vol. 7, No. 1: 1–12

A Spatial Post-Processing Algorithm for Images of Night Scenes

William B. Thompson and Peter Shirley
University of Utah

James A. Ferwerda
Cornell University

Abstract. The standard technique for making images viewed at daytime lighting levels look like images of night scenes is to use a low overall contrast, low overall brightness, desaturation, and to give the image a "blue shift." This paper introduces two other important effects associated with viewing real night scenes: visible noise, and the loss of acuity with little corresponding perceived blur.

1. Introduction

While many researchers have studied how to display images of daylit (*photopic*) scenes (a good overview can be found in [Durand, Dorsey 00]), little work has dealt with the display of images of night (*scotopic*) scenes. The work that has been done uses some combination of a reduction in contrast/brightness, desaturation, a blue shift, and a low pass filter [Upstill 95], [Tumblin, Rushmeier 93], [Ward 94], [Ferwerda et al. 96]. These strategies have also been used for many years in film [Samuelson 84].

There are two subjective properties of night viewing that are missing from previous techniques. The first is that although there is a loss-of-detail at night, there is not a sense of blurriness. The second is the somewhat noisy appearance of night scenes. Both of these effects have been only partially explored scientifically [Makous 90], [Field, Brady 97].

315

In this paper we present techniques for augmenting images with the loss-of-detail and noise effects associated with night vision. Because the subjective properties of scotopic effects are themselves not fully understood, and because we have the added difficulty of trying to mimic these effects in photopically-viewed images, we do not use a specific quantitative model to drive our method. Rather, we developed an empirical image processing technique that is consistent with what is known, and highlight the features of the techniques by demonstrating them on a test pattern and real images.

2. Background

While scotopic vision has not been studied to the extent as photopic vision, there is still a wealth of knowledge about it (e.g., see [Hess et al. 90]). We review some of that work here, and conclude that while it does not provide us with an explicit model to control our image processing, it does give us useful subjective information.

2.1. Visual Acuity and Perception of Blur

Visual acuity is reduced in scotopic viewing due, in part, to the relative sparsity of rods versus cones in the central retina. Subjectively, however, night does not appear blurry [Hess 90]. This has been observed in people who have only rod vision ("complete achromats") as well; Knut Nordby, a perceptual psychologist with that condition writes:

> I experience a visual world where things appear to me to be well-focused, have sharp and clearly defined boundaries and are not fuzzy or cloudy. [Nordby 90]

Simulating such loss of acuity by low-pass filtering the original image (e.g., [Ferwerda et al. 96]) fails to effectively capture the appearance of night scenes because when viewed in bright light, the images appear blurred rather than having the clearly defined boundaries Nordby describes.

Surprisingly little is known about the image cues that generate a sense of blur in the human vision system. While the power spectrum of natural images falls with frequency at about f^{-2}(e.g., [Ruderman 97]), there is substantial variability from image to image, even when the images appear sharp [Field, Brady 97]. As a result, a simple frequency distribution analysis will fail to predict the appearance of blur. Field and Brady argue that a person's sensitivity to blur depends on both the high frequency content of an image and the density of fine scale edges. In particular, an image will tend not to look blurred if fine detail edges are sufficiently crisp, independent of the number of fine scale edges in the image.

2.2. Noise

Noise in night scenes is also experienced subjectively. This is not surprising if one thinks of the scotopic vision system as a gain-control amplifier where low signals will have a noisy output [Lamb 90]. However, there are more potential sources of noise than this simple explanation suggests. These include the quantum number of photons, noise originating in the receptors themselves, and noise in the neural circuitry [Sharpe 90]. These various types of noise are themselves only partially quantified, and they are composed in complex ways which are not fully understood [Makous 90]. The noise is additive, but can behave as if multiplicative because neural noise might be added after logarithmic gain control [Makous 90].

3. Night Filtering

One way to simulate the loss of acuity in scotopic viewing would be to use some form of anisotropic diffusion (e.g., [Tumblin, Turk 99]), which has the effect of smoothing the geometry of edges without blurring across the edges. We have adopted a spatial filtering approach as an alternative because it does not require an iterative process and because the frequencies preserved can be more easily controlled. As noted above, an image will tend not to look blurred if fine detail edges are sufficiently crisp, independent of the number of fine scale edges in the image. For an original image I scaled to the range $[0.0 - 1.0]$, we start by low pass filtering the image, using convolution with a Gaussian kernel G_{blur}, where the standard deviation σ_{blur} is chosen to remove fine scale detail that would not be visible at night:

$$I_{\text{blur}} = G_{\text{blur}} * I. \tag{1}$$

It is not sufficient to simply apply a standard sharpening operator such as a narrow support unsharp mask to this blurred image, since there are no longer high frequencies to emphasize. A broader band sharpening filter, such as a larger extent unsharp mask, produces noticeable ringing. Instead, we apply a sharpening operator tuned to the finest detail edges remaining after the initial low pass filtering operation. This is done by first creating a bandpass-filtered image using the difference-of-Gaussian method, where the filter selects out the highest remaining spatial frequencies. The original image is convolved with a second Gaussian kernel G_{blur_2} with standard deviations $\sigma_{\text{blur}_2} = 1.6\,\sigma_{\text{blur}}$, after which we take the difference of the two blurred images:

$$I_{\text{blur}_2} = G_{\text{blur}_2} * I, \tag{2}$$

$$I_{\text{diff}} = I_{\text{blur}} - I_{\text{blur}_2}. \tag{3}$$

I_{diff} is a close approximation to the $\nabla^2 G$ function used in some edge detectors and is a bandpassed version of I [Marr, Hildreth 80]. I_{blur} can be decomposed into the bandpassed component plus an even lower pass component:

$$I_{\text{blur}} = I_{\text{blur}_2} + I_{\text{diff}}. \tag{4}$$

The appearance of blur can be substantially reduced while still attenuating fine detail by sharpening the edges in the bandpassed component. One way to do this would be to multiply I_{diff} by an appropriately chosen constant $\alpha > 1.0$. This is similar to a large-neighborhood unsharp mask and produces the same problems with ringing. Instead, we exploit the fact that edges are located at zero crossings of I_{diff} [Marr, Hildreth 80]. Edge sharpening can thus be accomplished by increasing the contrast of I_{diff} for values near 0:

$$I_{\text{night}} = I_{\text{blur}_2} + I_{\text{diff}}{}^{1.0/\gamma_{\text{edge}}} \quad \text{for} \quad \gamma_{\text{edge}} > 1.0. \tag{5}$$

The value of γ_{edge} affects the apparent crispness of the final edges. A value of 1.25 works well for a wide range of naturally occurring images.

The literature on dark noise is not sufficiently definitive as to allow the development of a precise physiologically or psychophysically based approach to transforming images so that they look more night-like, even when viewed in bright light. The problem is further compounded by the fact that different types of display devices can have very different effects on the appearance of high frequency image noise. As a result, we have adopted an approach in which we add zero-mean, uncorrelated Gaussian noise to images after they have been spatially filtered, with the standard deviation σ_{noise} of the noise adjusted subjectively depending on the display device ($\sigma_{\text{noise}} = .0125$ for the examples presented in this paper).

4. Implementation Issues

We assume the input is a displayable RGB image. If our source image is a high-dynamic range image, it should first be tone-mapped using one of the standard techniques (e.g., [Reinhard 02]). This RGB image is then mapped to a scotopic luminance image, where each pixel is a single number indicating the "brightness" of the pixel as seen at night. This will tend to favor blues because the rods are more sensitive to blues than to greens and reds. This can either be done heuristically using a linear weighted sum of the RGB channels or can be done by first converting to XYZ using one of the standard manipulations such as:

$$\begin{bmatrix} X \\ Y \\ Z \end{bmatrix} = \begin{bmatrix} 0.5149 & 0.3244 & 0.1607 \\ 0.2654 & 0.6704 & 0.0642 \\ 0.0248 & 0.1248 & 0.8504 \end{bmatrix} \begin{bmatrix} R \\ G \\ B \end{bmatrix}. \tag{6}$$

We can then use an empirical formula to approximate the scotopic luminance V [Larson et al. 97]:

$$V = Y \left[1.33 \left(1 + \frac{Y + Z}{X} \right) - 1.68 \right]. \tag{7}$$

This results in a single-channel image that has values in approximately the [0,4] range. That image should then be scaled and multiplied by a bluish grey to give an unfiltered "night image." Some authors suggest a blue with chromaticity approximately (0.03, 0.03) below the white point [Durand, Dorsey 00], although many films have a more saturated blue in their day-for-night scenes. Thus for each pixel we have

$$c_{\text{night}} = kV c_{\text{blue}}, \tag{8}$$

where k and c_{blue} are chosen empirically.

The resulting bluish image is then spatially processed using the methods discussed in Section 3. If some of the pixels are above the scotopic level, e.g., a car headlight or an area near a flame, then we can combine a day and a night image. If we consider the original RGB image to have a color c_{day} at each pixel, then we can use a scotopic fraction s at each pixel and blend the images pixel by pixel:

$$c = s c_{\text{night}} + (1 - s) c_{\text{day}}. \tag{9}$$

The fraction s can be chosen based on V, or can be chosen in a more principled manner (e.g., [Ferwerda et al. 96]).

5. Examples

We assume that the images we process have already undergone some day-for-night processing (reductions in brightness, contrast, and saturation, together with a blue shift). The images in this section were first transformed using a photo editing tool rather than a more formal tone-mapping algorithm. However, the techniques should be applicable to any initial day-for-night image.

Figure 1 illustrates the effect of applying the spatial filtering portion of the method to a test pattern consisting of gratings of different scales. σ_{blur} was chosen such that the individual bars in the center grouping would be resolvable, while the fine detail of the right-most bars would not. Figure 2 shows intensity plots of the test grating. The difference between night filtering

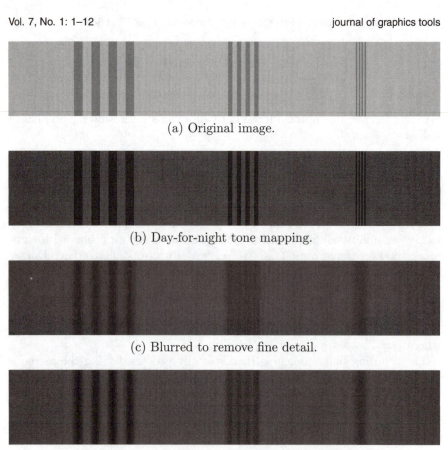

(a) Original image.

(b) Day-for-night tone mapping.

(c) Blurred to remove fine detail.

(d) Night-filtered to preserve same level of fine detail.

Figure 1. Test grating: The same level of detail is resolvable with Gaussian blurring and night filtering, but the night-filtered image looks sharper.

and simple blurring is most apparent when comparing the mid-sized bars in Figures 2(d) and 2(e). The night filtering does more than enhance the contrast: The definition of the bars relative to the overall contrast is significantly increased, which is why Figure 1(d) looks crisper than Figure 1(c).

Figures 3–8 illustrate the method on a natural image. Prior to spatial processing, the original image was corrected to account for photographic and digitization nonlinearities, with γ set based on a calibration target. Figure 4 was produced from Figure 3 by a manual day-for-night mapping involving a reduction in brightness, contrast, and saturation, as well as a blue shift. Figure 5 shows what would happen if the loss of acuity associated with viewing the scene at night was simulated by blurring. Figure 6 preserves the same level of detail as Figure 5, without appearing to be nearly as blurry. Figure 8 shows the effects of adding noise to Figure 6.

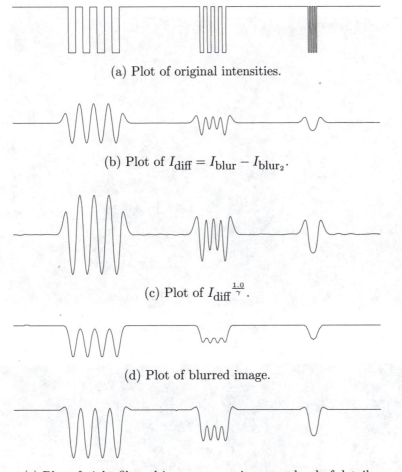

(a) Plot of original intensities.

(b) Plot of $I_{\text{diff}} = I_{\text{blur}} - I_{\text{blur}_2}$.

(c) Plot of $I_{\text{diff}}^{\frac{1.0}{\gamma}}$.

(d) Plot of blurred image.

(e) Plot of night-filtered image preserving same level of detail.

Figure 2. Plots of test grating images.

6.   Discussion

There are three parameters of the operator: σ_{blur} which is physically based (smallest resolvable detail), and γ_{edge} and σ_{noise} that are set subjectively. Experience suggests that a single setting is effective over a wide range of differing imagery. Because edges are stable over time in an image sequence, the edge operator we use will preserve locality and will also be stable [Marr, Hildreth 80]. Thus the operator is well-suited for animations. While the operator is designed for non-scientific applications, it might be calibrated to be useful in preference to a linear blur for low-vision simulations.

Figure 3. Original image. (See Color Plate XXVI.)

Figure 4. Day-for-night tone mapping. (See Color Plate XXVII.)

Figure 5. Blurred to remove fine detail. (See Color Plate XXVIII.)

Figure 6. Night-filtered, with the same level of fine detail as in Figure 5. (See Color Plate XXIX.)

Figure 7. Blurred plus noise. (See Color Plate XXX.)

Figure 8. Night filtering plus noise. (See Color Plate XXXI.)

Acknowledgements. Henrik Jensen, Daniel Kersten, Gordon Legge, Jack Loomis, and Simon Premoze provided helpful comments and discussion. This work was supported in part by National Science Foundation grants 89-20219, 95-23483, 98-18344 and 00-80999.

References

[Durand, Dorsey 00] F. Durand and J. Dorsey. "Interactive Tone Mapping." In *Eurographics Workshop on Rendering*, pp. 219–230. Wien: Springer-Verlag, 2000.

[Ferwerda et al. 96] J. A. Ferwerda, S. Pattanaik, P. S. Shirley, and D. P. Greenberg. "A Model of Visual Adaptation for Realistic Image Synthesis." In *Proceedings of SIGGRAPH 96, Computer Graphics Proceedings, Annual Conference Series*, edited by Holly Rushmeier, pp. 249–258, Reading, MA: Addison Wesley, 1996.

[Field, Brady 97] D. J. Field and N. Brady. "Visual Sensitivity, Blur and the Sources of Variability in the Amplitude Spectra of Natural Scenes." *Vision Research* 37(23): 3367–3383 (1997).

[Hess 90] R. F. Hess. "Rod-Mediated Vision: Role of Post-Receptorial Filters." In *Night Vision*, edited by R. F. Hess, L. T. Sharpe, and K. Nordby, pp. 3–48. Cambridge: Cambridge University Press, 1990.

[Hess et al. 90] R. F. Hess, L. T. Sharpe, and K. Nordby, eds. *Night Vision*. Cambridge: Cambridge University Press, 1990.

[Lamb 90] T. D. Lamb. "Dark Adaptation: A Re-Examination." In *Night Vision*, edited by R. F. Hess, L. T. Sharpe, and K. Nordby, pp. 177–222. Cambridge: Cambridge University Press, 1990.

[Larson et al. 97] G. W. Larson, H. Rushmeier, and C. Piatko. "A Visibility Matching Tone Reproduction Operator for High Dynamic Range Scenes." *IEEE Transactions on Visualization and Computer Graphics* 3(4): 291–306 (October–December 1997).

[Makous 90] W. Makous. "Absolute Sensitivity." In *Night Vision*, edited by R. F. Hess, L. T. Sharpe, and K. Nordby, pp. 146–176. Cambridge: Cambridge University Press, 1990.

[Marr, Hildreth 80] D. Marr and E. Hildreth. "Theory of Edge Detection." *Proc. Royal Society London* B 207: 187–217 (1980).

[Nordby 90] K. Nordby. "Vision in the Complete Achromat: A Personal Account." In *Night Vision*, edited by R. F. Hess, L. T. Sharpe, and K. Nordby, pp. 3–48. Cambridge: Cambridge University Press, 1990.

[Reinhard 02] E. Reinhard. "Parameter Estimation for Photographic Tone Repro-
    dution." *journal of graphics tools* 7(1): 45–52 (2002).

[Ruderman 97] D. L. Ruderman. "The Statistics of Natural Images." *Network:
    Computation in Neural Systems* 5(4): 517–548 (1997).

[Samuelson 84] D. W. Samuelson. *Motion Picture Camera Techniques, second edi-
    tion.* London: Focal Press, 1984.

[Sharpe 90] L. T. Sharpe. "The Light Adaptation of the Human Rod Vision Sys-
    tem." In *Night Vision*, edited by R. F. Hess, L. T. Sharpe, and K. Nordby,
    pp. 49–124. Cambridge: Cambridge University Press, 1990.

[Tumblin, Rushmeier 93] J. Tumblin and H. Rushmeier. "Tone Reproduction for
    Computer Generated Images." *IEEE Computer Graphics and Applications*
    13(6): 42–48 (November 93).

[Tumblin, Turk 99] J. Tumblin and G. Turk. "LCIS: A boundary hierarchy for
    detail-preserving contrast reduction." In *Proceedings of SIGGRAPH 99, Com-
    puter Graphics Proceeding, Annual Conference Series*, edited by Alyn Rock-
    wood, pp. 83–90, Reading MA: Addison Wesley Longman, 1999.

[Upstill 95] S Upstill. *The Realistic Presentation of Synthetic Images: Image
    Processing in Computer Graphics.* Ph. D. thesis, University of California
    at Berkeley, 1995.

[Ward 94] G. Ward. "A Contrast-Based Scalefactor for Luminance Display." In
    Graphics Gems IV, edited by P. Heckbert, pp. 415–421. Boston: Academic
    Press, 1994.

Web Information:

Images are available at
http://www.acm.org/jgt/papers/ThompsonShirleyFerwerda02

William B. Thompson, School of Computing, University of Utah, 50 S. Central
Campus Drive, Rm. 3190, Salt Lake City, UT 84112 (thompson@cs.utah.edu)

Peter Shirley, School of Computing, University of Utah, 50 S. Central Campus Drive,
Rm. 3190, Salt Lake City, UT 84112 (shirley@cs.utah.edu)

James A. Ferwerda, Computer Graphics Center, Cornell University, 580 Rhodes
Hall, Ithaca, NY 14853 (jaf2@cornell.edu)

Received April 1, 2002; accepted in revised form May 30, 2002.

Current Contact Information:

William B. Thompson, School of Computing, University of Utah, 50 S. Central Campus Drive, Rm. 3190, Salt Lake City, UT 84112 (thompson@cs.utah.edu)

Peter Shirley, School of Computing, University of Utah, 50 S. Central Campus Drive, Rm. 3190, Salt Lake City, UT 84112 (shirley@cs.utah.edu)

James A. Ferwerda, Computer Graphics Center, Cornell University, 580 Rhodes Hall, Ithaca, NY 14853 (jaf2@cornell.edu)

Vol. 8, No. 2: 17–30

Fast, Robust Image Registration for Compositing High Dynamic Range Photographs from Hand-Held Exposures

Greg Ward

Exponent – Failure Analysis Assoc.

Abstract.   In this paper, we present a fast, robust, and completely automatic method for translational alignment of hand-held photographs. The technique employs percentile threshold bitmaps to accelerate image operations and avoid problems with the varying exposure levels used in high dynamic range (HDR) photography. An image pyramid is constructed from grayscale versions of each exposure, and these are converted to bitmaps which are then aligned horizontally and vertically using inexpensive shift and difference operations over each image. The cost of the algorithm is linear with respect to the number of pixels and effectively independent of the maximum translation. A three million pixel exposure can be aligned in a fraction of a second on a contemporary microprocessor using this technique.

1.   Introduction

Although there are a few cameras entering the commercial market that are capable of direct capture of high dynamic range (HDR) photographs, most researchers still employ a standard digital or analog camera and composite HDR images from multiple exposures using the technique pioneered by Debevec and Malik [Debevec and Malik 97]. One of the chief limitations of this technique is its requirement that the camera be absolutely still between exposures. Even tripod mounting will sometimes allow slight shifts in the camera's position that yield blurry or double images. If there were a fast and reliable method to align image exposures, one would be able to take even handheld photos and reconstruct an HDR image. Since many of the midpriced digital cameras on the consumer market have a bracketed exposure mode, it is quite easy to take three to five exposures with different speed settings, suitable for HDR reconstruction. After taking 50 or so hand-held exposure sequences ourselves (each having 5 exposures), we made a number of unsuccessful attempts to reconstruct HDR images using available tools and methods, and decided to develop a new approach.

After experimenting with a few different algorithms, some borrowed from others and some developed ourselves, we found the bitmap method described here, which is very efficient at determining the optimal pixel-resolution alignment between exposures. The basic idea behind our approach is simple, and we explain this along with the more subtle points not to be neglected in Section 2. Some of the results and ideas for future work are presented in Section 3, followed by a brief conclusion.

2.   Method

Input to our alignment algorithm is a series of N 8-bit grayscale images, which may be approximated using only the green channel, or better approximated from 24-bit sRGB using the formula below.[1]

$$\text{grey} = (54 * \text{red} + 183 * \text{green} + 19 * \text{blue})/256$$

One of the N images is arbitrarily selected as the reference image, and the output of the algorithm is a series of $N - 1$ (x, y) integer offsets for each of the remaining images relative to this reference. These exposures may then be recombined efficiently into an HDR image using the camera response function, which may be computed using either Debevec and Malik's original SVD technique [Debevec and Malik 97], or as we have done in this paper, using the polynomial method of Mitsunaga and Nayar [Mitsunaga and Nayar 99].

We focused our computation on integer pixel offsets because they can be used to quickly recombine the exposures without resampling. Also, we found that 90% of the hand-held sequences we took did not require rotational alignment. Even in sequences where there was some discernable rotation, the effect of a good translational alignment was to push the blurred pixels out to the edges, where they are less distracting to the viewer. Rotation and subpixel alignment are interesting problems in their own right, and we discuss them briefly at the end of the paper.

---

[1]We have not found the red and blue channels to make any difference to the alignment results, so we just use the green channel.

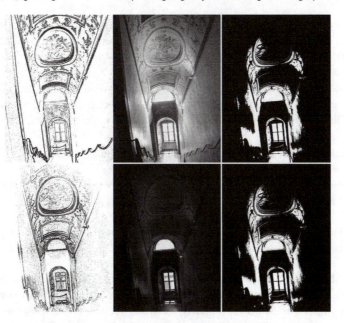

Figure 1. Two unaligned exposures (middle) and their corresponding edge bitmaps (left) and median threshold bitmaps (right). The edge bitmaps are not used in our algorithm, precisely because of their tendency to shift dramatically from one exposure level to another. In contrast, the MTB is stable with respect to exposure. (See Color Plate XXXII.)

We found out the hard way that conventional approaches to image alignment usually fail when applied to images with large exposure variations. In general, edge-detection filters are dependent on image exposure, as shown in the left side of Figure 1, where edges appear and disappear at different exposure levels. Edge-matching algorithms are therefore ill-suited to the exposure alignment problem. We tried the Skeleton Subspace Deformation (SSD) code of Hector Yee [Yee 01], which took a considerable length of time to find and align edges, and failed on the majority of our high-resolution examples. Similarly, we were frustrated in our efforts to find a point-feature matching algorithm that could reliably identify exposure-invariant features in our bracketed sequences [Cheng et al. 96].

The alternate approach we describe in this paper has the following desirable features:

- Alignment is done on bilevel images using fast, bit-manipulation routines
- Alignment is insensitive to image exposure
- Alignment includes noise filtering for robustness

The results of a typical alignment are shown in Figure 6 in the results section.

If we are to rely on operations such as moving, multiplying, and subtracting pixels over an entire high-resolution image, our algorithm is bound to be computationally expensive, unless our operations are very fast. Bitmap images allow us to operate on 32 or 64 pixels at a time using bitwise integer operations, which are very fast compared to byte-wise arithmetic. We experimented with a few different binary image representations until we found one that facilitated image alignment independent of exposure level: the *median threshold bitmap*.

A median threshold bitmap (MTB) is defined as follows:

1. Determine the median 8-bit value from a low-resolution histogram over the grayscale image pixels.

2. Create a bitmap image with 0s where the input pixels are less than or equal to the median value and 1s where the pixels are greater.

Figure 1 shows two exposures of an Italian stairwell in the middle and their corresponding edge maps on the left and MTBs on the right. In contrast to the edge maps, our MTBs are nearly identical for the two exposures. Taking the difference of these two bitmaps with an exclusive-or (XOR) operator shows where the two images are misaligned, and small adjustments in the x and y offsets yield predictable changes in this difference due to object coherence. However, this is not the case for the edge maps, which are noticeably different for the two exposures, even though we attempted to compensate for the camera response with an approximate response curve. Taking the difference of the two edge bitmaps would not give a good indication of where the edges are misaligned, and small changes in the x and y offsets yield unpredictable results, making gradient search problematic. More sophisticated methods to determine edge correspondence are necessary to use this information, and we can avoid these and their associated computational costs with our MTB-based technique.

The constancy of an MTB with respect to exposure is a very desirable property for determining image alignment. For most HDR reconstruction algorithms, the alignment step must be completed before the camera response can be determined, since the response function is derived from corresponding pixels in the different exposures. An HDR alignment algorithm that depends on the camera response function would thus create a chicken and egg problem. By its nature, an MTB is the same for any exposure within the usable range of the camera, regardless of the response curve. So long as the camera's response function is monotonic with respect to world radiance, the same scene will theoretically produce the same MTB at any exposure level. This is because the MTB partitions the pixels into two equal populations, one brighter and one darker than the scene's median value. Since the median value does not change in a static scene, our derived bitmaps likewise do not change with exposure level.[2]

There may be certain exposure pairs that are either too light or too dark to use the median value as a threshold without suffering from noise, and for these, we choose either the 17th or 83rd percentile as the threshold, respectively. Although our offset results are all relative to a designated reference exposure, we actually compute offsets between adjacent exposures, so the same threshold may be applied to both images. Choosing percentiles other than the 50th (median) results in fewer pixels to compare, and this makes the solution less stable, so we may choose to limit the maximum offset in certain cases. The behavior of percentile threshold bitmaps is otherwise the same as the MTB, including stability over different exposures. In the remainder of the paper, when we refer to the properties and operations of MTBs, one can assume that the same applies for other percentile threshold bitmaps as well.

Once we have our threshold bitmaps corresponding to the two exposures we wish to align, there are a number of ways to go about aligning them. One brute force approach is to test every offset within the allowed range, computing the XOR difference at each offset and taking the coordinate pair corresponding to the minimum difference. A more efficient approach might follow a gradient descent to a local minimum, computing only local bitmaps differences between the starting offset (0,0) and the nearest minimum. We choose a third method based on an image pyramid that is as fast as gradient descent in most cases, but is more likely to find the global minimum within the allowed offset range.

Multiscale techniques are well known in the computer vision and image-processing communities, and image pyramids are frequently used for registration and alignment. (See, for example, [Thevenaz et al. 98].) In our technique, we start by computing an image pyramid for each grayscale image exposure, with $\log_2(\texttt{max\_offset})$ levels past the base resolution. The resulting MTBs are shown for our two example exposures in Figure 2. For each smaller level in the pyramid, we take the previous grayscale image and filter it down by a factor of two in each dimension, computing the MTB from the grayscale result.[3]

To compute the overall offset for alignment, we start with the lowest resolution MTB pair and compute the minimum difference offset between them within a range of ±1 pixel in each dimension. At the next resolution level, we multiply this offset by 2 (corresponding to the change in resolution) and compute the minimum difference offset within a ±1 pixel range of this previ-

---

[2]Technically, the median value could change with changing boundaries as the camera moves, but such small changes in the median are usually swamped by noise, which is removed by our algorithm as we will explain.

[3]Be careful *not* to subsample the bitmaps themselves, as the result will be subtly different and could potentially cause the algorithm to fail.

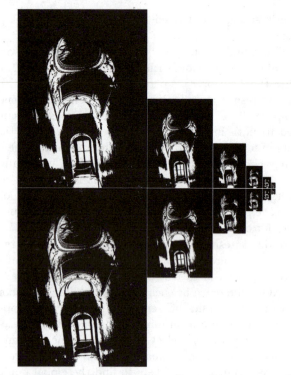

Figure 2. A pyramid of MTBs are used to align adjacent exposures one bit at a time. The smallest (right-most) image pair corresponds to the most significant bit in the final offset.

ous offset. This continues to the highest (original) resolution MTB, where we get our final offset result. Thus, each level in the pyramid corresponds to a binary bit in the computed offset value.

At each level, we need to compare exactly nine candidate MTB offsets, and the cost of this comparison is proportional to the size of the bitmaps. The total time required for alignment is thus linear with respect to the original image resolution and independent of the maximum offset, since our registration step is linear in the number of pixels, and the additional pixels in an image pyramid are determined by the size of the source image and the (fixed) height of the pyramid.

2.1. Threshold Noise

The algorithm just described works well in images that have a fairly bimodal brightness distribution, but can run into trouble for exposures that have a large number of pixels near the median value. In such cases, the noise in

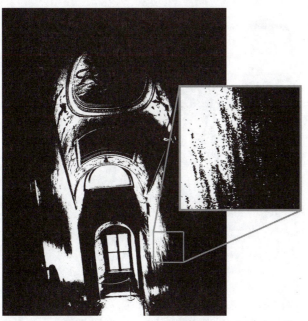

Figure 3. Close-up detail of noisy area of MTB in dark stairwell exposure (full resolution).

near-median pixels shows up as noise in the MTB, which destabilizes our difference computations.

The inset in Figure 3 shows a close-up of the pixels in our dark stairwell exposure MTB, and the kind of noise we see in some images. Computing the XOR difference between exposures with large areas like these yields noisy results that are unstable with respect to translation, because the pixels themselves tend to move around in different exposures. Fortunately, there is a straightforward solution to this problem.

Since our problem involves pixels whose values are close to our threshold, we can exclude these pixels from our difference calculation with an *exclusion bitmap*. Our exclusion bitmap consists of 0s wherever the grayscale value is within some specified distance of the threshold, and 1s elsewhere. The exclusion bitmap for the exposure in Figure 3 is shown in Figure 4, where we have zeroed all bits where pixels are within ±4 of the median value.

We compute an exclusion bitmap for each exposure at each resolution level in our pyramid, then take the XOR difference result for our candidate offset, ANDing it with both offset exclusion bitmaps to compute our final difference.[4]

---

[4]If we were to AND the exclusion bitmaps with the original MTBs before the XOR operation, we would inadvertently count disagreements about what was noise and what was not as actual pixel differences.

Figure 4. An exclusion bitmap, with zeroes (black) wherever pixels in our original image are within the noise tolerance of the median value.

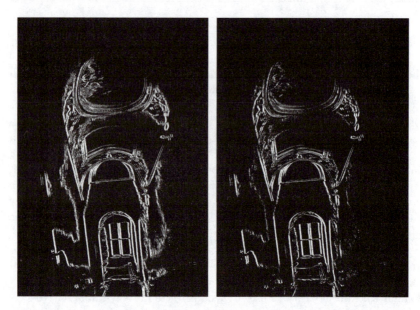

Figure 5. The original XOR difference of our unaligned exposures (left), and with the two exclusion bitmaps ANDed into the result to reduce noise in the comparison (right).

The effect is to disregard differences that are less than the noise tolerance in our images. This is illustrated in Figure 5, where we see the XOR difference of the unaligned exposures before and after applying the exclusion bitmaps. By removing those pixels that are close to the median, we clear the least reliable bit positions in the smooth gradients, but preserve the high confidence pixels near strong boundaries, such as the edges of the window and doorway. Empirically, we have found this optimization to be very effective in eliminating false minima in our offset search algorithm.

2.2. Overall Algorithm

The overall algorithm with the exclusion operator is given in the following recursive C function, GetExpShift. This function takes two exposure images, and determines how much to move the second exposure (img2) in x and y to align it with the first exposure (img1). The maximum number of bits in the final offsets is determined by the shift_bits parameter. We list the more important functions called by GetExpShift in Table 1, and leave their implementation as an exercise for the reader.

`ImageShrink2(const Image *img, Image *img_ret)`	Subsample the image img by a factor of two in each dimension and put the result into a newly allocated image img_ret.
`ComputeBitmaps(const Image *img, Bitmap *tb, Bitmap *eb)`	Allocate and compute the threshold bitmap tb and the exclusion bitmap eb for the image img. (The threshold and tolerance to use are included in the Image struct.)
`BitmapShift(const Bitmap *bm, int xo, int yo, Bitmap *bm_ret)`	Shift a bitmap by (xo,yo) and put the result into the preallocated bitmap bm_ret, clearing exposed border areas to zero.
`BitmapXOR(const Bitmap *bm1, constBitmap *bm2, Bitmap *bm_ret)`	Compute the "exclusive-or" of bm1 and bm2 and put the result into bm_ret.
`BitmapTotal(const Bitmap *bm)`	Compute the sum of all 1 bits in the bitmap.

Table 1. Functions called by GetExpShift.

```
GetExpShift(const Image *img1, const Image *img2, int shift_bits,
              int shift_ret[2])
{
    int          min_err;
    int          cur_shift[2];
    Bitmap       tb1, tb2;
    Bitmap       eb1, eb2;
    int    i, j;
    if (shift_bits > 0) {
        Image       sml_img1, sml_img2;
        ImageShrink2(img1, &sml_img1);
        ImageShrink2(img2, &sml_img2);
        GetExpShift(&sml_img1, &sml_img2, shift_bits-1, cur_shift);
        ImageFree(&sml_img1);
        ImageFree(&sml_img2);
        cur_shift[0] *= 2;
        cur_shift[1] *= 2;
    } else
        cur_shift[0] = cur_shift[1] = 0;
    ComputeBitmaps(img1, &tb1, &eb1);
    ComputeBitmaps(img2, &tb2, &eb2);
    min_err = img1->xres * img1->yres;
    for (i = -1; i <= 1; i++)
        for (j = -1; j <= 1; j++) {
            int          xs = cur_shift[0] + i;
            int          ys = cur_shift[1] + j;
            Bitmap       shifted_tb2;
            Bitmap       shifted_eb2;
            Bitmap       diff_b;
            int          err;
            BitmapNew(img1->xres, img1->yres, &shifted_tb2);
            BitmapNew(img1->xres, img1->yres, &shifted_eb2);
            BitmapNew(img1->xres, img1->yres, &diff_b);
            BitmapShift(&tb2, xs, ys, &shifted_tb2);
            BitmapShift(&eb2, xs, ys, &shifted_eb2);
            BitmapXOR(&tb1, &shifted_tb2, &diff_b);
            BitmapAND(&diff_b, &eb1, &diff_b);
            BitmapAND(&diff_b, &shifted_eb2, &diff_b);
            err = BitmapTotal(&diff_b);
            if (err < min_err) {
                shift_ret[0] = xs;
                shift_ret[1] = ys;
                min_err = err;
            }
            BitmapFree(&shifted_tb2);
            BitmapFree(&shifted_eb2);
        }
    BitmapFree(&tb1); BitmapFree(&eb1);
    BitmapFree(&tb2); BitmapFree(&eb2);
}
```

Computing the alignment offset between two adjacent exposures is simply a matter of calling the `GetExpShift` routine with the two image struct's (img1 and img2), which contain their respective threshold and tolerance values. (The threshold values must correspond to the same population percentiles in the two exposures.) We also specify the maximum number of bits allowed in our returned offset, shift_bits. The shift results computed and returned in shift_ret will thus be restricted to a range of $\pm 2^{\text{shift\_bits}}$.

There is only one subtle point in the above algorithm, which is what happens at the image boundaries. If we aren't careful, we might inadvertently shift nonzero bits into our candidate image, and count these as differences in the two exposures, which would be a mistake. That is why we need the `BitmapShift` function to shift 0s into the new image areas, so that applying the shifted exclusion bitmap to our XOR difference will clear these exposed edge pixels as well. This also explains why we need to limit the maximum shift offset, because if we allow this to be unbounded, then our lowest difference solution will also have the least pixels in common between the two exposures—one exposure will end up shifted completely off the other. In practice, we have found a shift_bits limit of 6 ($\pm$ 64 pixels) to work fairly well most of the time.

2.3. Efficiency Considerations

Clearly, the efficiency of our overall algorithm is going to depend on the efficiency of our bitmap operations, as we perform nine shift tests with six whole-image bitmap operations apiece. The `BitmapXOR` and `BitmapAND` operations are easy enough to implement, as we simply apply bitwise operations on 32-bit or 64-bit words, but the `BitmapShift` and `BitmapTotal` operators may not be so obvious.

For the `BitmapShift` operator, we start with the observation that any two-dimensional shift in a bitmap image can be reduced to a one-dimensional shift in the underlying bits, accompanied by a clear operation on one or two edges for the exposed borders. Implementing a one-dimensional shift of a bit array requires at most a left or right shift of B bits per word with a reassignment of the underlying word positions. Clearing the borders then requires clearing words where sequences of 32 or 64 bits are contiguous, and partial clearing of the remainder words. The overall cost of this operator, although greater than the XOR or AND operators, is still modest. Our own `BitmapShift` implementation includes an additional Boolean parameter that turns off border clearing. This allows us to optimize the shifting of the threshold bitmaps, which have their borders cleared later by the exclusion bitmap, and don't need the `BitmapShift` operator to clear them.

For the `BitmapTotal` operator, we precompute a table of 256 integers corresponding to the number of 1 bits in the binary values from 0 to 255 (i.e.,

0, 1, 1, 2, 1, 2, 2, 3, 1, ..., 8). We then break each word of our bitmap into byte-sized chunks, and add together the corresponding bit counts from the precomputed table. This results in a speed-up of at least eight times over counting individual bits, and may be further accelerated by special-case checking for zero words, which occur frequently in our application.

The author has implemented the above operations in a bitmap class using C++, which we are happy to make available to anyone who sends us a request for the source code.

3.   Results and Discussion

Figure 6 shows the results of applying our image alignment algorithm to all five exposures of the Italian stairwell, with detail close-ups showing before and after alignment. The misalignment shown is typical of a hand-held exposure sequence, requiring translation of several pixels on average to bring the exposures back atop each other. We have found that even tripod exposures sometimes need adjustment by a few pixels for optimal results.

We have applied our simple translational alignment algorithm to more than 100 hand-held exposure sequences. Overall, our success rate has been about 84%, with 10% giving unsatisfactory results due to image rotation. About 3% of our sequences failed due to excessive scene motion—usually waves or ripples on water that happened to be near the threshold value and moved between frames, and another 3% had too much high frequency content, which made the MTB correspondences unstable.[5] Most of the rotation failures were mild, leaving at least a portion of the HDR image well aligned. Other failures were more dramatic, throwing alignment off to the point where it was better not to apply any translation at all. It may be possible to detect such failure cases by looking at the statistics of MTB differences in successful and unsuccessful alignments, but we have not tested this idea as yet.

Although exposure rotation is less frequent and less objectionable than translation, it happens often enough that automatic correction is desirable. One possible approach is to apply our alignment algorithm separately to each quadrant of the image. If the computed translations are the same or nearly so, then no rotation is applied. If there are significant differences attributable to rotation, we could apply a rotation-filtering algorithm to arrive at an aligned result. This would be more expensive than straight image translation, but since it happens in relatively few sequences, the average cost might be acceptable. The divide-and-conquer approach might also enable us to detect inconsistent offset results, which would be a sign that our algorithm was not performing correctly.

---

[5]We thought we might correct some of these alignments if we allowed a ±2-pixel shift at each level of the pyramid rather than ±1, but our tests showed no improvement.

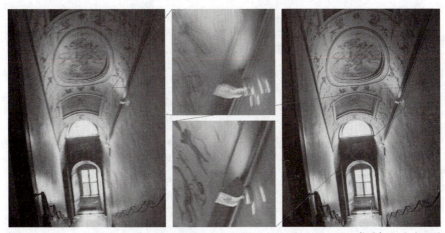

Figure 6. An HDR image composited from unaligned exposures (left) and detail (top center). Exposures aligned with our algorithm yield a superior composite (right) with clear details (bottom center). (See Color Plate XXXIII.)

If we are willing to accept the filtering expense associated with rotational alignment, we may as well align images at a subpixel level. To accomplish this, the MTB algorithm could be extended so that one or two levels above the highest resolution bitmap are generated via bilinear or bicubic interpolation on the original image. Such a representation would be needed for rotation anyway, so the additional expense of this approach would be nominal.

Ultimately, we would like digital camera manufacturers to incorporate HDR capture into their products. Although the algorithm we describe has potential for onboard image alignment due to its speed and small memory footprint, it would be better if the exposures were taken close enough together that camera motion was negligible. The real problem is in establishing a compressed, high dynamic range image format that the camera manufacturers and imaging software companies can support. We have proposed two standards for encoding high dynamic range colors [Ward 91], [Larson 98], but work still needs to be done to incorporate these color spaces into an image compression standard, such as JPEG 2000 [Christopoulos et al. 00].

References

[Cheng et al. 96] Yong-Qing Cheng, Victor Wu, Robert T. Collins, Allen R. Hanson, and Edward M. Riseman. "Maximum-Weight Bipartite Matching Technique and Its Application in Image Feature Matching." In *SPIE Conference on Visual Communication and Image Processing*. Bellingham, WA: SPIE, 1996.

[Christopoulos et al. 00] C. Christopoulos, A. Skodras, and T. Ebrahimi. "The JPEG2000 Still Image Coding: An Overview." *IEEE Transactions on Consumer Electronics* 46:4 (2000), 1103–1127.

[Debevec and Malik 97] Paul Debevec and Jitendra Malik. "Recovering High Dynamic Range Radiance Maps from Photographs." In *Proceedings of SIGGRAPH 97, Computer Graphics Proceedings, Annual Conference Series*, edited by Turner Whitted, pp. 27–34, Reading, MA; Addison-Wesley, 1997.

[Larson 98] Greg Ward Larson. "LogLuv Encoding for Full-Gamut, High-Dynamic Range Images." *journal of graphics tools* 3:1 (1998), 15–31.

[Mitsunaga and Nayar 99] T. Mitsunaga and S. K. Nayar. "Radiometric Self Calibration." In *Proceedings of IEEE Conference on Computer Vision and Pattern Recognition*, pp. 1472–1479. Los Alamitos: IEEE Press, 1999.

[Thevenaz et al. 98] P. Thevenaz, U. E. Ruttimann, and M. Unser. "A Pyramid Approach to Subpixel Registration Based on Intensity." *IEEE Transactions on Image Processing* 7:1 (1998), 27–41.

[Ward 91] Greg Ward. "Real Pixels." In *Graphics Gems II*, edited by James Arvo, pp. 80–83. San Diego: Academic Press, 1991.

[Yee 01] Hector Yee. Personal communication, 2001.

Web Information:

http://www.acm.org/jgt/papers/Ward03.

Greg Ward, Anyhere Software, 1200 Dartmouth St., Apt. C, Albany, CA 94706-2358, http://www.anyhere.com (gward@lmi.net)

Received September 19, 2002; accepted in revised form May 15, 2003.

New Since Original Publication

Since this paper was published, we have made one minor change to the algorithm that makes it more robust. This improvement was inspired by a comment from one of the anonymous referees, who suggested there would be cases where our pyramid descent algorithm would "set the wrong bit" higher up in the pyramid and not be able to correct the mistake the rest of the way down. This usually occurs when the difference between the errors associated with two possible alignment offsets are equal or nearly so, forcing a somewhat arbitrary choice at that level. If the choice turns out to be wrong, subsequent bits may end up being all zero or all one, trying to get closer to the correct offset. Such a mistake would result in an alignment error of at least one pixel, and probably more. In general, a mistake at a higher level could lead the system to a divergent local minimum—one far removed from the global minimum we seek.

We arrived at a simple extension that allows for half-pixel offsets to be returned by the GetExpShift() function. This is not done by shifting the bitmaps by a half pixel, which would be difficult, but by noting when neighboring offsets produce nearly identical error sums. Rather than tracking the minimum error offset in our search loop, we record the error sums for each candidate offset, then sort these results from the smallest to largest error. We average together all offsets that are within some tolerance of the smallest error. (We use 0.025% of the total pixels at this level.) We round this average off to the nearest half-pixel position, which then becomes a full pixel at the next higher resolution pyramid. There, we center our next search. For the final result, we ignore any half-pixel offsets, though a better result might be obtained from using this extra precision in a bicubic filter.

By itself, the half-pixel modification inoculates us against choosing a wrong bit value during our descent, and costs us virtually nothing. To be even more certain of avoiding local minima, we extended our search to encompass the closest 21 offsets to each level's starting point, rather than the closest nine as in the original algorithm. We spend a little over twice as long in our search with the larger radius, but it allows us to recover alignment on an additional 4–8% of our exposure sequences that were previously failing. Since the algorithm is in general very fast, we felt it was a reasonable trade. (A five exposure sequence from a 4 megapixel camera takes less than six seconds to align on a 1.8-GHz G5 processor.)

Updated Web Information:

A software implementation of this algorithm is currently available for download at www.anyhere.com.

Current Contact Information:

Greg Ward, 1200 Dartmouth St., #C, Albany, CA 94706 (gward@lmi.net)

journal of graphics tools
a forum to share hands-on techniques with your peers

jgt includes:

- Tricks and hacks
- Innovative techniques and algorithms
- Experience and advice
- Production notes
- Novel research ideas
- Surveys
- Tutorials

Contribute!
Submit your article online at
http://journaltool.akpeters.com

Editor-in-Chief
Ronen Barzel

Consulting Editor
David Salesin

Founding Editor
Andrew Glassner

Editorial Board
Tomas Akenine-Möller
Richard Chuang
Eric Haines
Chris Hecker
John Hughes
Darwyn Peachey
Peter Shirley
Paul Strauss
Wolfgang Stürzlinger

The *journal of graphics tools* is a quarterly journal whose primary mission is to provide the computer graphics research, development, and production communit with practical ideas and techniques that solve real problems. We aim to bridge th gap between new research ideas and their use as tools by the computer graphics professional.

Annual subscription rate - Volume 10 (2005)
$160.00 institutional; $70.00 individual
Shipping
In the U.S.: $7.50;
in Canada: $15.00;
all other countries: $25.00

A K Peters, Ltd.
888 Worcester St. Suite 230
Wellesley, MA 02482
USA

Tel 781-416-2888
Fax 781-416-2889
Email service@akpeters.com
Web www.akpeters.com